AF356133

Structure and Properties of Aluminium Alloys

Structure and Properties of Aluminium Alloys

Editor

Franc Zupanič

MDPI • Basel • Beijing • Wuhan • Barcelona • Belgrade • Manchester • Tokyo • Cluj • Tianjin

Editor
Franc Zupanič
University of Maribor
Slovenia

Editorial Office
MDPI
St. Alban-Anlage 66
4052 Basel, Switzerland

This is a reprint of articles from the Special Issue published online in the open access journal *Metals* (ISSN 2075-4701) (available at: https://www.mdpi.com/journal/metals/special_issues/structure_properties_aluminium_alloys).

For citation purposes, cite each article independently as indicated on the article page online and as indicated below:

LastName, A.A.; LastName, B.B.; LastName, C.C. Article Title. *Journal Name* **Year**, *Volume Number*, Page Range.

ISBN 978-3-0365-1008-8 (Hbk)
ISBN 978-3-0365-1009-5 (PDF)

Cover image courtesy of Franc Zupanič.

Contents

About the Editor

Franc Zupanič is a lecturer in the field of Materials Science at the University of Maribor, Faculty of Mechanical Engineering. He has published several textbooks in this field. His main research topics are quasicrystals in Al-alloys, high-strength and heat-resistant aluminium alloys, continuous casting of nickel and cobalt superalloys, solidification and heat treatment. He has given his full attention to materials characterization, especially metallography and mechanical testing using indentation. He has also formed strong collaboratations with industry.

Preface to "Structure and Properties of Aluminium Alloys"

Structure and properties of aluminium alloys. https://www.mdpi.com/journal/metals/special_issues/structure_properties_aluminium_alloys

The annual world production of aluminum and aluminum alloys has been increasing in recent decades, reaching more than 70 million tons in 2020. The future of this industry is bright, as the applications of Al and its alloys have strongly diversified in automotive, aerospace, building, and other industries. Aluminum's main property is its low density, and, more importantly, very high specific properties compared to other metallic and nonmetallic materials. The properties of aluminum alloys can contribute to a significant decrease in energy consumption and CO_2 emissions, especially in transportation. The main prerequisite for the future success of aluminum and its alloys is further improvements in existing alloys, and the development of novel aluminum alloys. In addition to conventional fabrication methods (casting, forming, powder metallurgy), additive manufacturing technologies enable further tailoring of alloys' microstructure and new combinations of properties. The properties of aluminum alloys are mainly based on their structure, from the atomic scale to the macrostructure, as seen by the naked eye. This book focuses on the relationships between the manufacturing, structure and properties of aluminum alloys. The papers presented give an account of the scientific and technological state-of-the-art of aluminum alloys in 2020.

Franc Zupanič
Editor

Article

Microstructure and Properties of a Novel Al-Mg-Si Alloy AA 6086

Franc Zupanič [1,*], Matej Steinacher [2], Sandi Žist [2] and Tonica Bončina [1]

[1]   Faculty of Mechanical Engineering, University of Maribor, SI-2000 Maribor, Slovenia; tonica.boncina@um.si
[2]   Impol d.d., SI-2310 Slovenska Bistrica, Slovenia; matej.steinacher@impol.si (M.S.); sandi.zist@impol.si (S.Ž.)
*   Correspondence: franc.zupanic@um.si; Tel.: +386-2-220-7863

Abstract: In this work, we investigated a novel Al-Mg-Si alloy, which was developed from an AA 6082, in order to considerably improve the yield and tensile strengths whilst retain excellent ductility. The new alloy possesses a higher content of Si than specified by AA 6082, and, in addition, it contains copper and zirconium. The alloy was characterized in the as-cast condition, after homogenization, extrusion, and T6 heat treatment using light microscopy, scanning and transmission electron microscopy with energy dispersive spectrometry, X-ray diffraction, differential thermal analysis and tensile testing. After T6 temper, tensile strengths were around 490 MPa with more than 10% elongation at fracture. The microstructure consisted of small-grained Al-rich matrix with α-AlMnSi and Al_3Zr dispersoids, and Q'-AlCuMgSi and β-Mg_2Si-type precipitates.

Keywords: aluminium; ageing; microstructure; tensile strength

Citation: Zupanič, F.; Steinacher, M.; Žist, S.; Bončina, T. Microstructure and Properties of a Novel Al-Mg-Si Alloy AA 6086. *Metals* **2021**, *11*, 368. https://doi.org/10.3390/met11020368

Academic Editor: Marcello Cabibbo

Received: 30 January 2021
Accepted: 19 February 2021
Published: 23 February 2021

Publisher's Note: MDPI stays neutral with regard to jurisdictional claims in published maps and institutional affiliations.

1. Introduction

The manufacturing of green vehicles, characterized by lower fuel consumptions and CO_2 emissions, is a must for the automotive industry. This can be achieved through a reduction of the vehicle weight by replacing heavier steel parts with lighter, high-strength and, at the same time, ductile aluminium alloys. Al-Mg-Si alloys (series 6xxx), in particular, can be used [1]. Some of them are used for parts that require excellent crash performance [2]. There are also attempts to improve the mechanical properties of Al-Si-Mg alloys [3–5]. Table 1 gives tensile properties of some Al-alloys, which shows that the required properties of AA 6610A are highest among Al-Mg-Si alloys.

Table 1. The required mechanical properties of 6xxx alloys in T6 temper according to EN 755-2 standard.

Alloy	$R_{p0.2}$/MPa	R_m/MPa	A/%
AA 6082	min. 260	min. 310	min. 8
AA 6182	min. 330	min. 360	min. 9
AA 6110A	min. 380	min. 410	min. 10

Al-Mg-Si alloys are typical precipitation hardening alloys. The strengthening effect in Al-Mg-Si alloys is achieved through β' and β'' precipitates [6–8]. These precipitates are metastable variants of the equilibrium Mg_2Si phase. They turn to Mg_2Si when the alloy is held prolonged at higher temperatures (between the ageing and solvus temperatures). One obtains a combination of 6xxx and 2xxx alloy by the addition of copper to Al-Mn-Si. In these alloys, the phase Q-AlCuMgSi dominates but not Θ-Al_2Cu that is predominant in 2xxx alloys. During ageing, Q' precipitates are formed in the α-Al, which strengthen the matrix in addition to β'' precipitates [9–11].

Zirconium can decrease the size of crystal grains during solidification. However, zirconium in the alloy can reduce the efficiency of Al-Ti-B grain refiners due to the poisoning

1

of TiB_2 and Al_3Ti heterogeneous nucleation sites [12]. It is essential that zirconium form Al_3Zr dispersoids at higher temperatures [13–16]. These dispersoids considerably decrease the tendency to crystal growth during heat treatment and hot plastic deformation. In this way, zirconium contributes to strengthening of the alloys by dispersion hardening and hardening by grain boundaries. Namely, the latter hardening is essential when the crystal grains are fine [17,18].

This study's main aim was to produce a new high-strength Al-Mg-Si alloy under industrial conditions (semi-continuous casting, homogenization, extrusion and T6 temper), to determine its microstructure after each step of the manufacturing process and its tensile properties.

2. Materials and Methods

The alloy AA 6086 [19] was developed from the AA 6082. The main differences are in the contents of Si, Cu and Zr. The AA 6082 contains 0.7–1.3 wt.% Si, while the AA 6086 contains 1.3–1.7 wt.% Si. The higher silicon content can contribute to a higher fraction of Mg_2Si-based precipitates in the microstructure. The increase in Cu-content from 0.1 to 0.8 wt.% Cu (0–0.1 wt.% Cu in AA 6082) can provoke Cu-rich precipitates' formation, which can further increase the precipitation hardening effect. The addition of Zr in the range from 0.15–0.25 wt.% Zr can promote the formation of Al_3Zr dispersoids, which can reduce grain growth during homogenization and solution treatment and increase grain strengthening. One of the possible primary applications of this alloy is to produce automobile steering tie rods, which require high-strength and ductility.

The 20 t charge of the alloy was prepared in an induction furnace (Otto Junker GmbH, Simmerath-Lammersdorf, Germany) at a temperature of 760 °C. The zirconium was added with the master alloy AlZr10. The melt was poured into a holding furnace (Otto Junker GmbH, Simmerath-Lammersdorf, Germany) for 90 min at temperatures between 730 in 740 °C. The billets with a diameter of 282 mm were semi-continuously cast using the AirSlip Casting Technology (Wagstaff Inc, Spokane Valley, WA 99216 USA). AlTi5B1 grain refiner was added during casting. The chemical composition of the alloy is given in Table 2. The billets were homogenized at 530 °C for 5 h and then extruded on 55 MN direct extrusion press (GIA Clecim Press, Albacete, Spain) into bars with a diameter of 57 mm. The extrusion rate was 7 m/min and extrusion ratio 4.95. After extrusion, the bars were solution treated at 530 °C for 2 h and then artificially aged at 180 °C for 8 h (T6 temper).

Table 2. The chemical composition of the alloy AA 6086 in wt.%. as determined by ICP-OES (Thermo ARL 4460 Optical Emission Spectrometer, ThermoFisher Scientific, Waltham, MA, USA).

Alloy	Al	Si	Fe	Cu	Mn	Mg	Cr	Zn	Ti	Zr
AA 6082	Remain	0.92	0.26	0.07	0.56	0.70	0.14	0.03	0.02	0.01
AA 6086	Remain	1.58	0.18	0.55	0.71	1.01	0.19	0.17	0.04	0.17

The differential thermal analysis (DTA) was carried out using Mettler Toledo 851 (Mettler Toledo, Columbus, Ohio 43240, USA). A sample was heated and cooled at a rate of 10 °C/min. The samples in the as-cast condition were examined metallographically using a light microscope Nikon EPIPHOT 300 (Tokyo, Japan), and a scanning electron microscopes (SEM) Sirion 400 NC (FEI, Eindhoven, Netherlands). In SEM, we also carried out microchemical analysis using energy dispersive spectroscopy—EDS (Oxford Analytical, Bicester, UK). Samples for grain analysis were prepared on Struers Lectropol (Struers, Ballerup, Denmark) using Barker's etchant (50 mL HBF_4 (48%) + 950 mL H_2O at 25 V for 90 s). The analysis was performed under the $50\times$ magnification according to a standard ASTM E112-intercept method. X-ray diffraction was done at synchrotron Elettra (Trieste, Italy), using X-rays with a wavelength of 0.099996 nm. Thermo-Calc software (Thermo-Calc software AB, Solna, Sweden) and database TCAL5 were used to simulate solidification according to the Scheil model [20] and calculate equilibrium phases as a function of temperature.

Lamellae for transmission electron microscopy (TEM) were prepared using a focused ion beam (FIB by FEI, Helios) for the as-cast and heat-treated specimens. High-resolution TEM (HRTEM), and energy-dispersive X-ray spectroscopy (EDXS) were carried out in an FEI Titan 80–300 (FEI, Eindhoven, Netherlands) image corrected electron microscope.

Four tensile specimens of alloy AA 6086 and seventeen of alloy AA 6082 were tested. They were taken at $D/4$ in the longitudinal direction. The diameter of the specimens was 20 mm and the gauge length 100 mm. The tensile tests were done by Zwick Z250 (Zwick-Roell Testing Systems GmbH, Fürstenfeld, Austria) according to ISO 6892-1 standard.

3. Results

3.1. Phase Identification in the As-Cast Condition and after Homogenization

Figure 1 gives the calculated phase composition and the effect of Zr on the liquidus temperature. The modelled liquidus temperature is around 700 °C when the alloy contains 0.15% Zr and rises to about 740 °C at 0.25% Zr. Thus, the alloy can be melted entirely in typical industrial furnaces, which can attain temperatures slightly above 750 °C. On the other hand, the effect on the solidus temperature is rather low, which remains faintly higher than 550 °C at Zr-contents up to 0.4%.

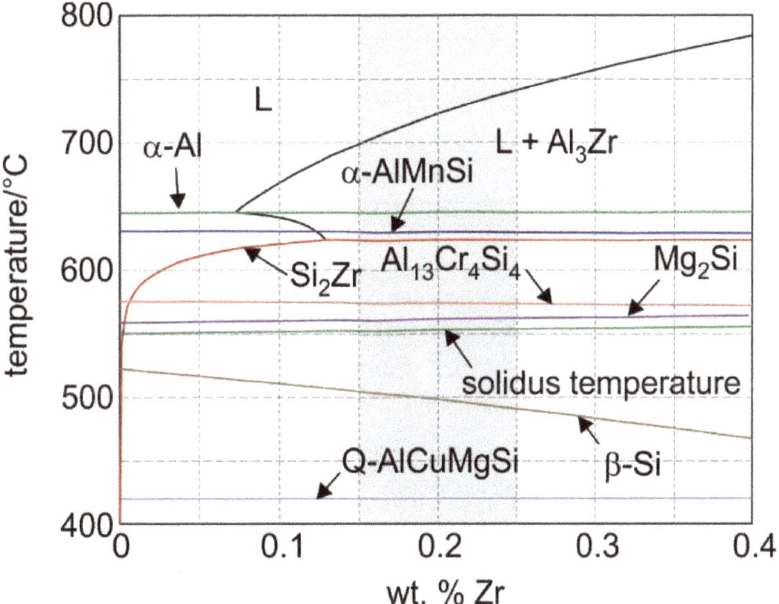

Figure 1. The effect of Zr on the liquidus temperature and formation of phases in the Al-corner. The shaded area indicates the content range of Zr in AA 6086, the contents of other elements are the same as in Table 2.

DTA-analysis (Figure 2) showed that the incipient melting point was at approximately 540 °C, which is more than 10 °C lower than the calculated solidus temperature. This is the result of non-equilibrium solidification. Scheil simulation showed that during solidification also Θ-Al$_2$Cu, β-Si and Q-AlCuMgSi can form. Figure 1 shows that β-Si and Q-AlCuMgSi can form in the solid-state, and Θ-Al$_2$Cu is not predicted as a stable phase. DTA-analysis also suggests that the homogenization temperature should be below the incipient meltings point below 540 °C.

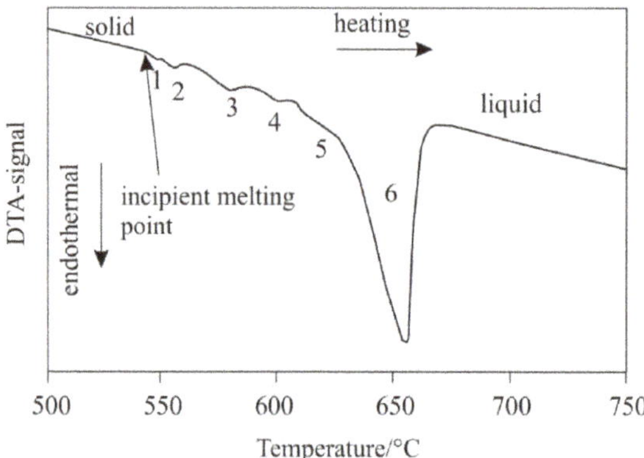

Figure 2. A section of the heating DTA-curve between 500 °C and 750 °C (the heating rate 10 °C/min, argon) showing six peaks occurring during the melting of the alloy, and that the incipient melting point lies approximately at 540 °C.

Figure 3 shows the light micrographs revealing the grain structure in the as-cast state and after homogenization treatment. The matrix α-Al solid solution possessed equiaxed dendritic grains with sizes between 180 and 215 µm. The grain structures were somewhat similar in both conditions.

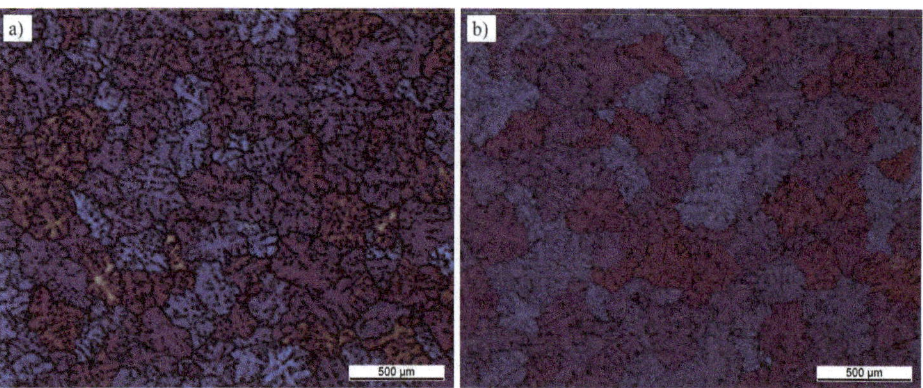

Figure 3. The light micrographs of the alloy AA 6086 in (**a**) the as-cast condition and (**b**) after homogenization treatment.

SEM-micrographs (Figure 4) revealed the presence of other phases in the interdendritic regions. The EDS-analysis showed that α-Al contained from 96.1 to 97.7 wt.% Al. At the centre of α-Al, the contents of Si and Cr were higher than in the interdendritic regions, the contents of Mg and Mn was less at the centre, while the contents of Cu and Zr were almost the same. The distribution of elements was uniform after homogenization. The contents of Si and Cr in the α-Al increased, the contents of Mg and Mn decreased, while almost no differences were observed in the Zr-content. X-ray diffraction revealed that the lattice parameters of the face-centred cubic FCC α-Al in both conditions were nearly the same ($a = 0.40458 \pm 0.00025$ nm in the as-cast state and $a = 0.40475 \pm 0.00020$ nm after homogenization, 0.004 nm smaller than in pure Al).

The X-ray diffraction patterns contained many peaks that often overlapped with one another. A detailed X-ray diffraction (XRD) analysis (Figure 5) revealed seven phases in the as-cast condition: α-Al, $MgSi_2$, tetragonal Al_3Zr, α-AlMnSi, $ZrSi_2$, Θ-Al_2Cu and β-Si and four after α-Al, $MgSi_2$, tetragonal Al_3Zr, α-AlMnSi, $ZrSi_2$, indicating that Θ-Al_2Cu and β-Si were dissolved. Nevertheless, the diffraction peaks remained at the same positions, because α-AlMnSi have a very high number of diffractions peaks at lower angles due to its very large unit cell. The relative height of some peaks changed, indicating that some phases partly or entirely dissolved.

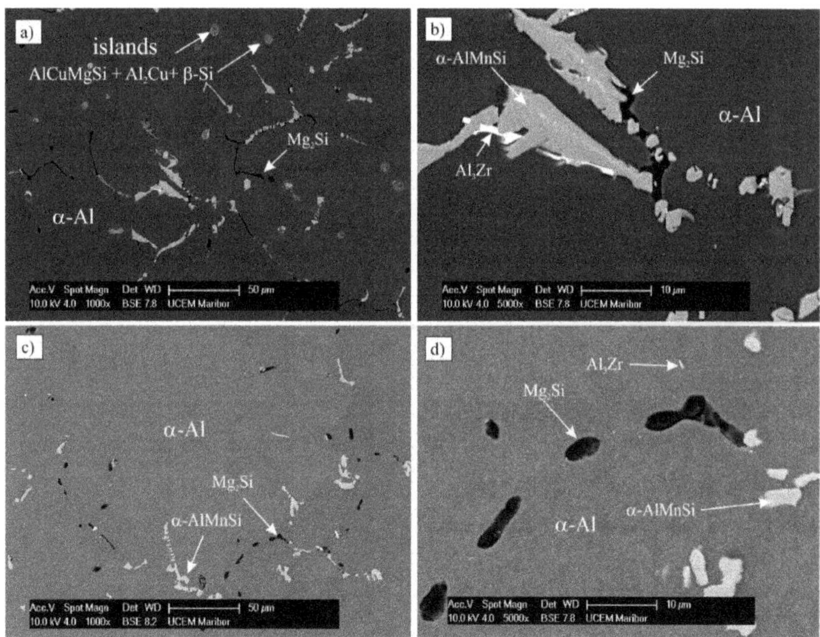

Figure 4. The backscattered electron micrographs (SEM) of the alloy AA 6086 in the as-cast condition (**a,b**) and after homogenization treatment (**c,d**).

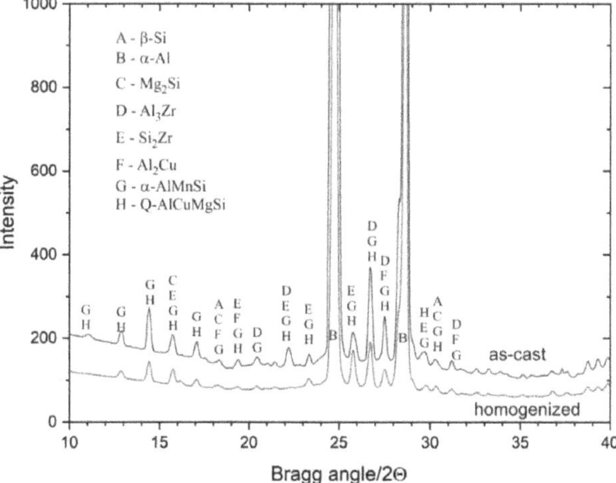

Figure 5. X-ray diffraction patterns of AA 6086 in the as-cast condition and after homogenization treatment. The peak positions of all phases found in the microstructure are indicated, showing strong overlapping of peaks.

Among the intermetallic phases, the volume fraction of the cubic α-AlMnSi was highest. EDS analysis confirmed that this phase also contained Fe and Cr. Typical chemical composition was 11–12 at.% Si; 2–5 at.% Cr, 11.8–12.6 at.% Mn, 8–10 at.% Fe, 0.5 at.% Cu and traces of Zr. The phase α-AlMnSi was present in the interdendritic and intergranular spaces. It has the same composition in both conditions, so the lattice parameters were almost the same (α-Al(Mn,Fe)Si: as-cast a = 1.26165 ± 0.00410 nm and homogenized a = 1.26287 ± 0.00397 nm). The size of α-AlMnSi did not change significantly after homogenization. However, the particles became more compact, with rounded edges. The calculation of equilibrium phases (Figure 6) shows that α-AlMnSi is relatively stable in this alloy. Only a small fraction of α-AlMnSi dissolves when heated to the homogenization temperature. The particles coarsen slowly to decrease the interfacial energy. It also indicates that the alloying elements in α-AlMnSi cannot be dissolved in α-Al and contribute to dispersion and precipitation hardening.

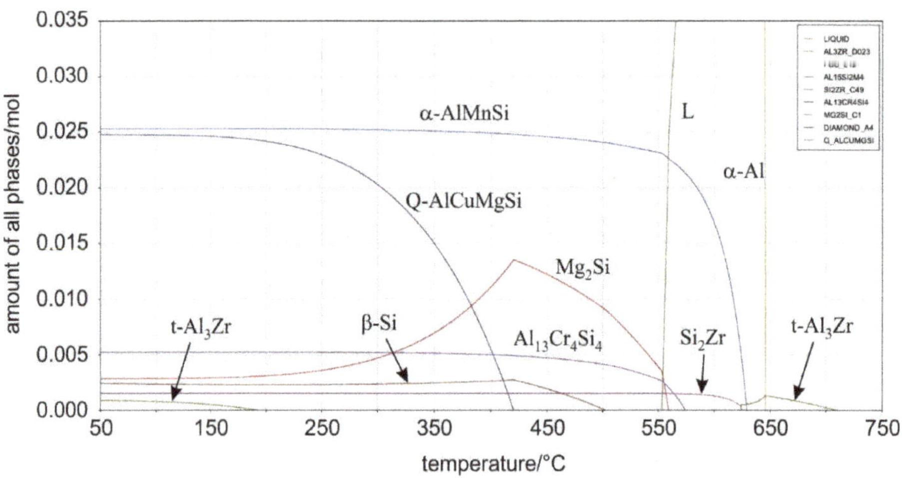

Figure 6. The calculated equilibrium phase fractions as a function of temperature. The calculated equilibrium solidus temperature T_s was slightly above 750 °C. The Al$_{13}$Cr$_4$Si$_4$ phase was not observed in the microstructure, as chromium did not form its own phase, but it was incorporated into α-AlMnSi.

The black Mg$_2$Si had a shape of Chinese script within a two-phase (α-Al + Mg$_2$Si) structure in the as-cast condition. During non-equilibrium solidification, an excess of Mg$_2$Si forms due to Mg and Si segregation in the interdendritic region. The calculation showed (Figure 6) that Mg$_2$Si should dissolve in a more significant amount at homogenization temperature. The diffusivities of Mg and Si in α-Al are high [21], so the non-equilibrium amount of Mg$_2$Si can dissolve at the homogenization temperature and distribute in entire α-Al grains. It was shown that the volume fraction of Mg$_2$Si dropped from 1.4% to 0.3% after homogenization, and the remaining particles significantly change their shape. They also became spheroidal in shape to degrease the interfacial energy. The dissolved Mg and Si are critical for the precipitation of β-type precipitates during artificial ageing. The lattice parameters were similar in both conditions (as-cast a = 0.63483 ± 0.00360 nm and homogenized a = 0.63422 ± 0.00126 nm).

Zirconium was present in platelike particles of the tetragonal Al_3Zr and within the orthorhombic Si_2Zr. Si_2Zr crystallized on the Al_3Zr platelets and was not separated from Al_3Zr. The closer microstructure analysis revealed that β-Si, Al_2Cu and Q-AlCuMgSi were present within the islands formed via a multiphase reaction at the terminal stages of solidification (Figure 4a). The Scheil solidification model predicted all discovered phases [20]. Phases β-Si, Al_2Cu and Q-AlCuMgSi present within the islands in the as-cast condition, dissolved completely during homogenization, which is in accordance with calculations (Figures 1 and 6).

Within α-Al, copious nucleation of dispersoids took place during homogenization (Figure 7). The TEM-investigation that α-AlMnSi and Al_3Zr dispersoids formed within the α-Al matrix during homogenization. Their sizes were between 100 and 500 nm, and had different shapes. They formed because of the supersaturation of the matrix with Zr and Mn after solidification. A detailed EDS-analysis showed that the most of dispersoids can be attributed to α-AlMnSi phase. The brighter phases in the BSE-SEM images, as well as in the HAADF TEM micrographs were identified as Al_3Zr dispersoids.

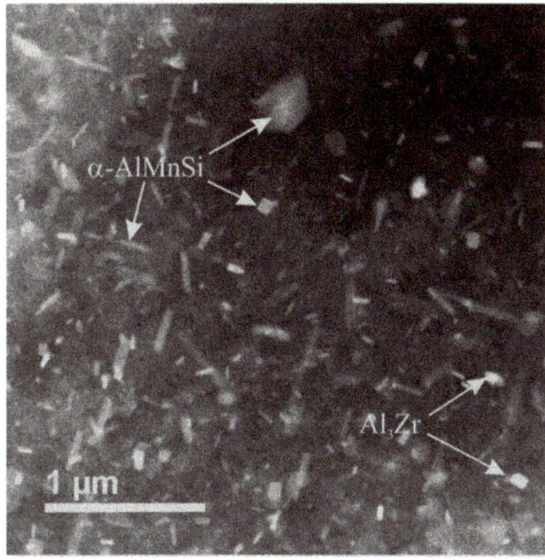

Figure 7. Dispersoids in α-Al after homogenization. The High-Angle Annular Dark Field (HAADF) image in TEM.

3.2. Microstructure of AA 6086 in the T6 Temper

The final microstructure and properties were achieved by the T6 temper, which consisted of the solution annealing and artificial ageing. Prior to this, the alloy was hot extruded, which oriented crystal grains in the pressing direction (length around 500 μm, thickness about 40 μm. Figure 8 shows a TEM-micrograph, which reveals the subgrain structure and dispersoids. During solution annealing, the crystal grains remain almost the same as after extrusion, but the subgrains became equiaxed (Figure 9). Since the subgrain sizes were only 2–5 μm in size, dispersoids effectively prevented excessive grain growth during solution treatment. Extensive EDS-analysis also revealed two types of dispersoids, which were identified before: α-AlMnSi and Al_3Zr.

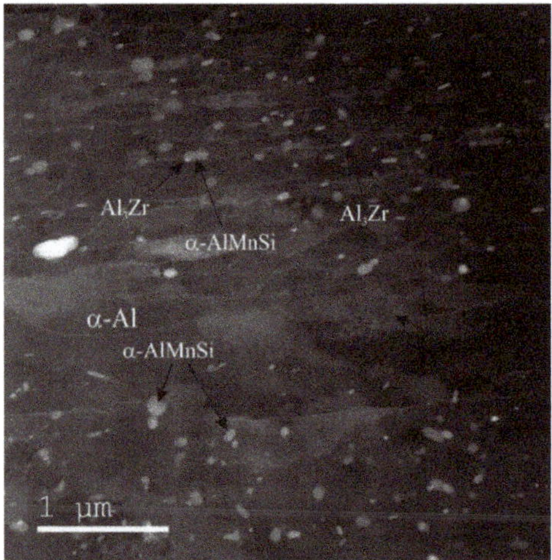

Figure 8. The microstructure after extrusion. The extrusion direction is from left to right. The High-Angle Annular Dark Field (HAADF) image in TEM.

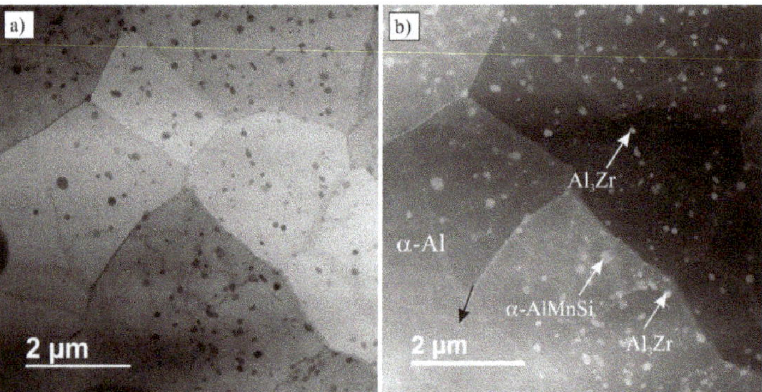

Figure 9. The transmission electron micrographs of the alloy after T6 temper, shoving subgrains and the distribution of dispersoids. (**a**) Bright-field micrograph and (**b**) HAADF micrograph.

Figure 10 showed the analysis of an α-AlMnSi dispersoid. EDS-showed that is also contained Cr and Fe. This particle had the z-axis parallel to the z-axis of Al, however, in the plane x-y no matching was found between these two phases.

Figure 11 showed the analysis of Al_3Zr dispersoid. EDS showed that it also contained silicon and a smaller amount of Ti, so it is correct to write $(Al,Si)_3Zr$ [17]. This particle possessed and excellent matching with the α-Al, which can be shown by HR-TEM micrograph (Figure 11c) and FFF of HR-TEM (Figure 11d). It is inferred that this particle had cubic L_{12}, which usually has the cube-to-cube orientation relationship with α-Al [22]. In addition to spheroidal Al_3Zr with $L1_2$ structure, platelike Al_3Zr particles were present. These particles had the same composition, but according to other works, they probably had a tetragonal structure [14].

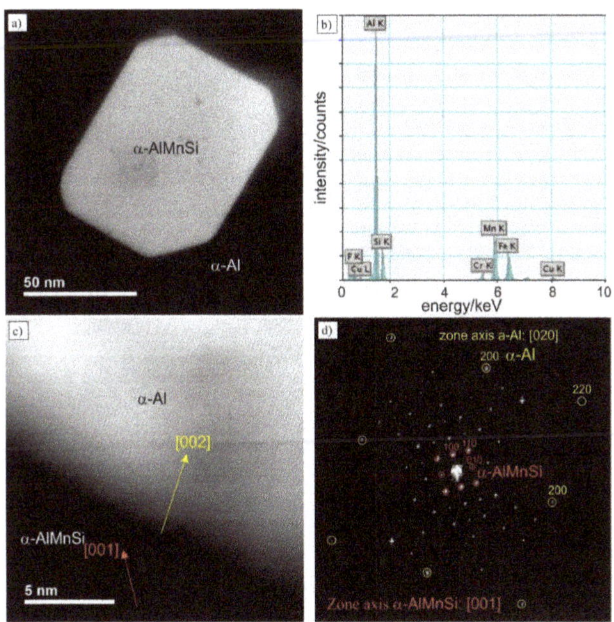

Figure 10. An α-AlMnSi dispersoid in the alloy after T6 temper. (**a**) HAADF micrograph, (**b**) EDS-spectrum, (**c**) High-resolution HAADF micrograph, (**d**) Fast-Fourier transform (FFT) of image (**c**). Zone axis α-Al: [001], zone axis of α-AlMnSi: [001].

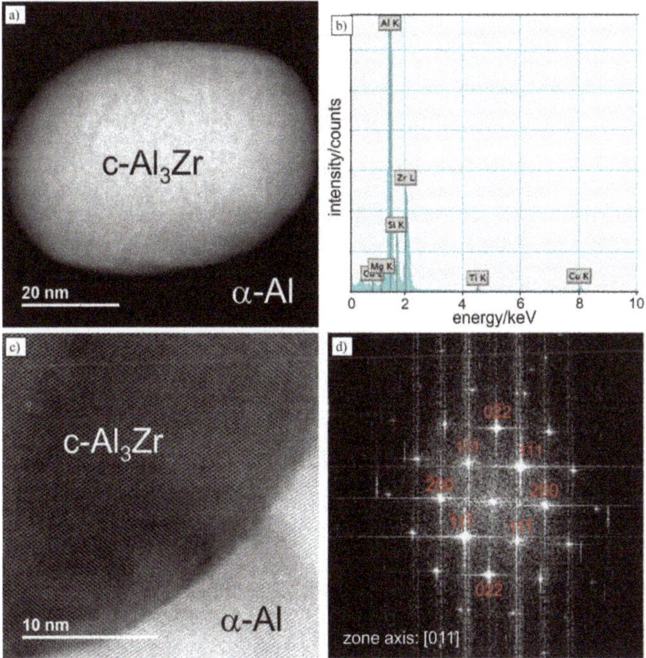

Figure 11. An c-Al$_3$Zr dispersoid in the alloy after T6 temper. (**a**) HAADF micrograph, (**b**) EDS-spectrum, (**c**) High-resolution HAADF micrograph, (**d**) FFT of image (**c**). Zone axis of c-Al$_3$Zr: [011].

Higher magnification showed a very high density of precipitates, which were mainly elongated β'-precipitates (Mg_2Si) and Q'-precipitates with different shapes (AlCuMgSi) (Figure 12). Figure 13 shows an individual Q'-precipitate in the α-Al, EDS-showed that it contained Al, Cu, Mg and Si.

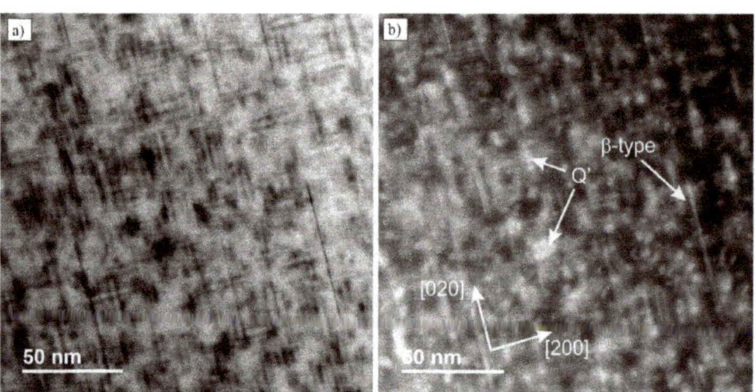

Figure 12. The transmission electron micrographs of the alloy after T6 temper, shoving precipitates. (**a**) Bright-field micrograph and (**b**) HAADF micrograph.

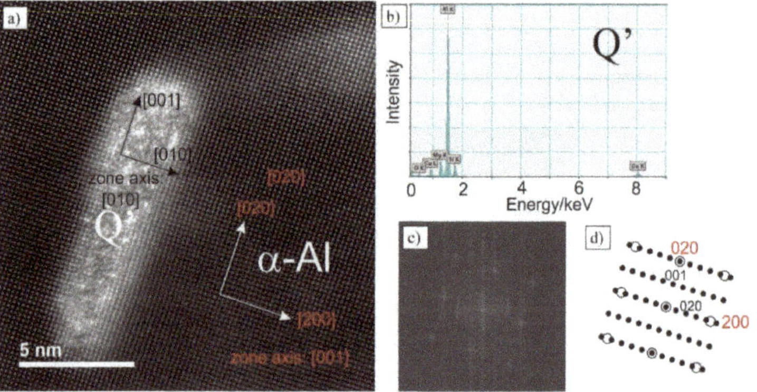

Figure 13. A Q'-precipitate. (**a**) The high resolution TEM micrograph of the Q'- precipitate in α-Al, (**b**) EDS-spectrum, (**c**) FFT of the HR-TEM micrograph, (**d**) model of FFT (Open circles: α-Al, black dots: Q'-precipitate.

3.3. Comparison of Tensile Properties of AA 6086 and AA 6082

The new alloy AA 6086 was developed by modifying the composition of AA 6082 to obtain higher strength. Table 3 compares the results of several tensile tests of both alloy and Figure 14 shows their typical stress-strain diagrams. The modification of chemical composition accompanied by typical manufacturing processes produced an increase of both 0.2 proof stress ($R_{p0.2}$) and tensile strength (R_m) for about 100 MPa, while the ductility did not change a lot.

Table 3. Comparison of tensile properties of AA 6082 (average values and standard deviations for 17 samples) and AA 6086 (average values and standard deviations for 4 samples).

Alloy	Sample	$R_{p0.2}$/MPa	R_m/MPa	A/%
AA 6082	#1	365	389	13.7
	#2	358	382	11.8
	#3	354	376	13.2
	#4	366	390	13.6
	#5	359	384	14
	#6	348	375	12.1
	#7	362	386	13.5
	#8	361	386	10.5
	#9	358	384	13.1
graph	#10	360	385	13
	#11	359	384	14.2
	#12	358	385	12
	#13	359	385	11.5
	#14	363	387	9.8
	#15	377	401	12.8
	#16	371	395	11.3
	#17	373	396	13.3
Average and standard deviation		361.8 ± 6.8	386.5 ± 6.3	12.5 ± 1.2
AA 6086	#1	450	490	12.3
graph	#2	456	488	9.7
	#3	461	492	12.4
	#4	464	495	11.8
Average and standard deviation		457.8 ± 5.3	491.2 ± 2.6	11.5 ± 1.1

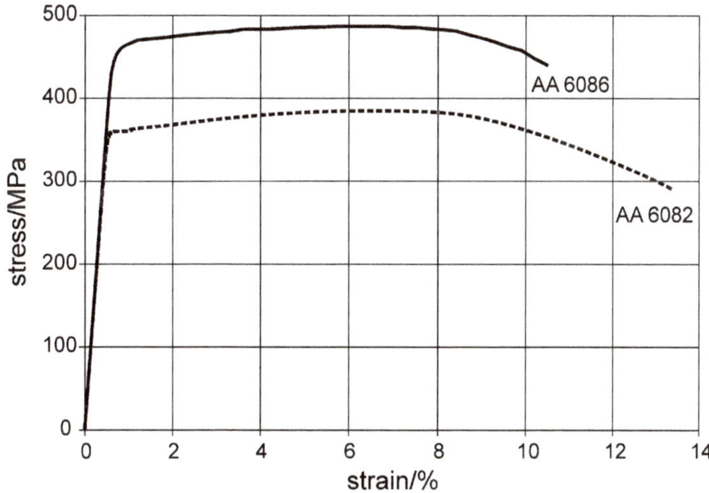

Figure 14. Typical tensile diagrams of the alloys AA 6082 ($R_{p0,2}$ = 360 MPa, R_m = 385 MPa, A = 13.0%, blue) and AA 6086 ($R_{p0,2}$ = 456 MPa, R_m = 488 MPa, A = 9.7%, red) in the T6 temper.

4. Discussion

By developing a novel alloy, the complete chemical composition of the AA 6082 was optimized [19]. The addition of Zr in the range 0.15–0.25% increased the liquidus temperature from about 650 °C to 700–740 °C, which requires sufficient superheating of the melt during melting, holding and casting to prevent formation of primary Al$_3$Zr. The particles can namely settle to the bottom of the furnace, and makes the Zr unavailable for formation of dispersoids in later manufacturing steps. It is also desirable that the cooling rates during solidification are sufficiently high to avoid formation of coarse Al_3Zr crystals. Al_3Zr reacts with the liquid via a peritectic reaction, thus forming α-Al with a higher Zr-content. During further cooling α-AlMnSi forms via a eutectic reaction. The α-AlMnSi particles are rather coarse and prone to changes upon subsequent homogenization and solution treatment. Therefore, these particles can be detrimental for ductility. Mg_2Si is formed during further cooling. Despite it does not dissolve completely during homogenization, it obtains the spheroidal shape, which is not so harmful for the ductility. Other phases that form during the terminal stages of solidification completely dissolve during homogenization. Upon homogenization also dispersoids form. The alloying elements that are required for formation of dispersoids does not come from the dissolving phases formed during solidification, but rather from the α-Al matrix that was supersaturated with rather slowly diffusing Mn and Zr. Predominantly forms α-AlMnSi and Al_3Zr. In the first stage, c-Al_3Zr forms [14], that was also identified in the investigated alloy. Cubic Al_3Zr can latter transform to tetragonal Al_3Zr. Plastic deformation with extrusion cause elongation of crystal grains in the extrusion direction. However, it also caused formation of elongated subgrains, which have only few micrometres in length and about 0.5 μm in thickness. The solution treatment caused change of subgrains from directional to equlaxed, with sizes 2–5 μm. The dispersoids present in the matrix effectively prevented dynamic recovery and recrystallization during hot extrusion and excessive subgrain growth during solution annealing. Thus, the small subgrains can significantly contribute to hardening [17,18].

Artificial ageing caused the formation of β-type precipitates (Mg_2Si), which are the primary hardening precipitates in the Al-Mn-Si alloys. The addition of copper caused the formation of Cu-rich precipitates, e.g., Q'-AlCuMgSi, which were also found in other cases, when Cu was added to 6XXX alloys [11].

The combined effect of different hardening mechanisms produces a rather strong alloy. The yield strength was 452.5–463.1 MPa, the tensile strength 488.6–493.8 MPa, while elongation at fracture was 9.6–12.6%. The strength levels strongly exceed the strengths of other 6xxx alloys. However, the ductility was decreased, probably due to the coarse particles of α-AlMnSi phase. Therefore, additional work is required to further improve the properties, and determine impact toughness, fatigue strength and corrosion resistance of this alloy.

Table 4 gives the presence of precipitates and dispersoids in some Al-alloys. The basic and well-known alloy AA 6082 is strengthened mainly by β-type precipitates and contains α-AlMnSi dispersoids. The newer alloy AA 6182 has almost the same composition but contains 0.05–0.2 wt.% Zr. This increases the tensile strength by 50 MPa. It should be stressed that the alloy AA 6082 reached the tensile strength of 386 MPa (Table 3), indicating that the alloy AA 6182 could attain strengths above 400 MPa. The alloy AA 6110A, which contains almost the same amount of copper as AA 6086, but no Zr, attains tensile strength 410 MPa. However, using the similar manufacturing route as in the current investigated AA 6086 alloy, tensile properties around 450 MPa could be achieved. The novel alloy AA 6086 should contain more β-type precipitates, more Mn-dispersoids and, in addition, Zr-dispersoids. Altogether can contribute to higher strength levels in this alloy. The alloy AA 6086 is not yet included in EN 755-2:2016 standard, but its likely minimum tensile strength would be 450 MPa.

Table 4. The presence of precipitates and dispersoids in selected Al-alloys, and minimum tensile strengths R_m in T6 temper according to EN 755-2:2016 (* possible minimum value for AA 6086, which is not yet included in EN 755-2:2016; -: not present, +: present, ++: present in a higher amount).

Alloy	β-Type	Q'-Precipitate	Mn-Dispersoids	Zr-Dispersoids	min. R_m/MPa
AA 6082	+	-	+	-	310
AA 6182	+	-	+	+	360
AA 6110A	+	+	+	-	410
AA 6086	++	+	++	+	450 *

The investigated AA 6086 efficiently uses advantages of other high-strength Al-Mg-Si alloys, and is a result of evolution, and not revolution. It should be stated that for attaining the tensile strengths up to 500 MPa in Al-Mg-Si alloys no additional strengthening mechanisms are required. It should be stressed that in the industry, even classical manufacturing processes are tightly monitored and controlled, which allows obtaining the properties above the minimum levels stated by standards.

One of the examples of the drastic increase of mechanical properties with a slight change of composition was given by Han et al. [23] who studied microstructure and mechanical properties of Al-Mg-Si-Cu alloy with high manganese content. They achieved more than 50% increase in the yield strength of the examined alloys with high manganese content compared to the commercial AA 6061. They stated that additional Mn-dispersoids were generating the strong dispersion hardening effect.

5. Conclusions

The investigation results showed that a slight modification of the chemical composition of a standard aluminium alloy and its manufacturing by rather conventional manufacturing process steps could result in a much higher strength (about 100 MPa) with a minute decrease of ductility. Nevertheless, some improvements are still possible by optimization of processing steps.

The addition of copper resulted in the formation of Q'-AlCuMgSi precipitates, which combined with β-type (Mg_2Si) precipitates contributed to strong precipitation hardening effect. The addition of Zr resulted in the formation of different Al_3Zr-type dispersoids, which, together with α-AlMnSi prevented excessive grain growth during homogenization and solution treatment.

This work indicates that somewhat small changes can lead to significant variations in properties. This comprehension my lead to the development of other aluminium alloys with improved properties, which are highly required in today's world.

6. Patents

The results of this investigation were significant by the application of the following patent issued by European Patent Office: EP3214191A1: A high-strength Al-Mg-Si aluminium alloy and its manufacturing process, Peter Cvahte, Marina Jelen, Dragojevič Vukašin, Steinacher Matej, Impol d. d., Slovenska Bistrica, Slovenia

Author Contributions: Conceptualization, T.B., F.Z., M.S.; methodology, F.Z., M.S.; formal analysis, T.B., M.S., F.Z.; investigation, T.B., M.S. and S.Ž.; writing—original draft preparation, F.Z., T.B.; writing—review and editing, F.Z., T.B., S.Ž.; visualization, T.B. and M.S.; supervision, T.B. All authors have read and agreed to the published version of the manuscript.

Funding: The XRD investigations were carried out at Elettra, Sincrotrone Trieste, Italy, in the framework of Proposal 20180085. The TEM investigations at TU Graz were enabled by the ESTEEM 2 Project, which has received funding from the European Union's Seventh framework program under Grant Agreement No. 312483. The authors acknowledge the financial support from the Slovenian Research Agency (research core funding No. P2-0120 and I0-0029) and by the Ministry of Education, Science and Sport, Republic of Slovenia (Program MARTINA, OP20.00369).

Data Availability Statement: Not applicable.

Acknowledgments: We acknowledge Luisa Barba for support by XRD experiments at Elettra, Christian Gspan for S/TEM investigations.

Conflicts of Interest: The authors declare no conflict of interest.

References

1. McQueen, H.J.; Celliers, O.C. Application of hot workability studies to extrusion processing. 3. Physical and mechanical metallurgy of Al-Mg-Si and Al-Zn-Mg alloys. *Can. Met. Q.* **1997**, *36*, 73–86. [CrossRef]
2. Schiffl, A.; Schiffl, I.; Hartmann, M.; Brötz, S.; Österreicher, J.; Kühlein, W. Analysis of Impact Factors on Crash Performance of High Strength 6082 Alloys Consider Extrudability and Small Modifications of the Profile Geometry. *Mater. Today Proc.* **2019**, *10*, 193–200. [CrossRef]
3. Hu, J.; Zhang, W.; Fu, D.; Teng, J.; Zhang, H. Improvement of the mechanical properties of Al–Mg–Si alloys with nano-scale precipitates after repetitive continuous extrusion forming and T8 tempering. *J. Mater. Res. Technol.* **2019**, *8*, 5950–5960. [CrossRef]
4. Wang, X.F.; Shi, T.Y.; Wang, H.B.; Zhou, S.Z.; Xie, C.; Wang, Y.G. Mechanical behavior and microstructure evolution of Al-Mg-Si-Cu alloy undertensile loading at different strain rates. *Mater. Res. Express* **2019**, *6*, 15. [CrossRef]
5. Sekhar, A.P.; Mandal, A.B.; Das, D. Mechanical properties and corrosion behavior of artificially aged Al-Mg-Si alloy. *J. Mater. Res. Technol.-JMRT* **2020**, *9*, 1005–1024. [CrossRef]
6. Farh, H.; Djemmal, K.; Guemini, R.; Serradj, F. Nucleation of dispersoids study in some Al-Mg-Si alloys. *Ann. Chim.-Sci. Mater.* **2010**, *35*, 283–289. [CrossRef]
7. Mao, H.; Kong, Y.; Cai, D.; Yang, M.; Peng, Y.; Zeng, Y.; Zhang, G.; Shuai, X.; Huang, Q.; Li, K.; et al. β″ needle-shape precipitate formation in Al-Mg-Si alloy: Phase field simulation and experimental verification. *Comput. Mater. Sci.* **2020**, *184*, 109878. [CrossRef]
8. Yang, M.; Chen, H.; Orekhov, A.; Lu, Q.; Lan, X.; Li, K.; Zhang, S.; Song, M.; Kong, Y.; Schryvers, D.; et al. Quantified contribution of β″ and β′ precipitates to the strengthening of an aged Al–Mg–Si alloy. *Mater. Sci. Eng. A* **2020**, *774*, 138776. [CrossRef]
9. Chakrabarti, D.J.; Laughlin, D.E. Phase relations and precipitation in Al–Mg–Si alloys with Cu additions. *Prog. Mater. Sci.* **2004**, *49*, 389–410. [CrossRef]
10. Sunde, J.K.; Marioara, C.D.; van Helvoort, A.T.J.; Holmestad, R. The evolution of precipitate crystal structures in an Al-Mg-Si(-Cu) alloy studied by a combined HAADF-STEM and SPED approach. *Mater. Charact.* **2018**, *142*, 458–469. [CrossRef]
11. Sunde, J.K.; Marioara, C.D.; Holmestad, R. The effect of low Cu additions on precipitate crystal structures in overaged Al-Mg-Si(-Cu) alloys. *Mater. Charact.* **2020**, *160*, 110087. [CrossRef]
12. Wang, Y.; Fang, C.M.; Zhou, L.; Hashimoto, T.; Zhou, X.; Ramasse, Q.M.; Fan, Z. Mechanism for Zr poisoning of Al-Ti-B based grain refiners. *Acta Mater.* **2019**, *164*, 428–439. [CrossRef]
13. Babaniaris, S.; Ramajayam, M.; Jiang, L.; Varma, R.; Langan, T.; Dorin, T. Effect of Al3(Sc,Zr) dispersoids on the hot deformation behaviour of 6xxx-series alloys: A physically based constitutive model. *Mater. Sci. Eng. A* **2020**, *793*, 139873. [CrossRef]
14. Meng, Y.; Cui, J.; Zhao, Z.; He, L. Effect of Zr on microstructures and mechanical properties of an AlMgSiCuCr alloy prepared by low frequency electromagnetic casting. *Mater. Charact.* **2014**, *92*, 138–148. [CrossRef]
15. Vlach, M.; Cizek, J.; Smola, B.; Melikhova, O.; Vlcek, M.; Kodetova, V.; Kudrnova, H.; Hruska, P. Heat treatment and age hardening of Al-Si-Mg-Mn commercial alloy with addition of Sc and Zr. *Mater. Charact.* **2017**, *129*, 1–8. [CrossRef]
16. Fu, J.N.; Yang, Z.; Deng, Y.L.; Wu, Y.F.; Lu, J.Q. Influence of Zr addition on precipitation evolution and performance of Al-Mg-Si alloy conductor. *Mater. Charact.* **2020**, *159*, 7. [CrossRef]
17. Himuro, Y.; Koyama, K.; Bekki, Y. Precipitation Behaviour of Zirconium Compounds in Zr-Bearing Al-Mg-Si Alloy. *Mater. Sci. Forum* **2006**, *519–521*, 501–506. [CrossRef]
18. Li, L.; Zhang, Y.; Esling, C.; Jiang, H.; Zhao, Z.; Zuo, Y.; Cui, J. Crystallographic features of the primary Al$_3$Zr phase in as-cast Al-1.36wt% Zr alloy. *J. Cryst. Growth* **2011**, *316*, 172–176. [CrossRef]
19. Cvahte, P.; Jelen, M.; Dragojevič, V.; Steinacher, M. A High-Strength Al-Mg-Si Aluminium Alloy and Its Manufacturing Process. European Patent Office EP3214191A1, 19 August 2020.
20. Zupanič, F.; Steinacher, M.; Bončina, T. As-cast microstructure of a novel Al-Mg-Si alloy. *Livar. Vestn.* **2019**, *66*, 125–136.
21. Du, Y.; Chang, Y.A.; Huang, B.Y.; Gong, W.P.; Jin, Z.P.; Xu, H.H.; Yuan, Z.H.; Liu, Y.; He, Y.H.; Xie, F.Y. Diffusion coefficients of some solutes in fcc and liquid Al: Critical evaluation and correlation. *Mater. Sci. Eng. A-Struct. Mater. Prop. Microstruct. Process.* **2003**, *363*, 140–151. [CrossRef]
22. Zupanič, F.; Gspan, C.; Burja, J.; Bončina, T. Quasicrystalline and L12 precipitates in a microalloyed Al-Mn-Cu alloy. *Mater. Today Commun.* **2020**, *22*, 100809. [CrossRef]
23. Han, Y.; Ma, K.; Li, L.; Chen, W.; Nagaumi, H. Study on microstructure and mechanical properties of Al–Mg–Si–Cu alloy with high manganese content. *Mater. Des.* **2012**, *39*, 418–424. [CrossRef]

Article

Bimodal Microstructure Obtained by Rapid Solidification to Improve the Mechanical and Corrosion Properties of Aluminum Alloys at Elevated Temperature

Irena Paulin [1,*], Črtomir Donik [1], Peter Cvahte [2] and Matjaž Godec [1]

[1]   Institute of Metals and Technology, Lepi pot 11, 1000 Ljubljana, Slovenia; crtomir.donik@imt.si (Č.D.);
     matjaz.godec@imt.si (M.G.)
[2]   IMPOL 2000 d.d., Partizanska ulica 38, 2310 Slovenska Bistrica, Slovenia; peter.cvahte@impol.si
*   Correspondence: irena.paulin@imt.si

Abstract: The demand for aluminum alloys is increasing, as are the demands for higher strength, with the aim of using lighter products for a greener environment. To achieve high-strength, corrosion-resistant aluminum alloys, the melt is rapidly solidified using the melt-spinning technique to form ribbons, which are then plastically consolidated by extrusion at elevated temperature. Different chemical compositions, based on adding the transition-metal elements Mn and Fe, were employed to remain within the limits of the standard chemical composition of the AA5083 alloy. The samples were systematically studied using light microscopy, scanning electron, and transmission microscopy with electron diffraction spectrometry for the micro-chemical analyses. Tensile tests and Vickers microhardness were applied for mechanical analyses, and corrosion tests were performed in a comparison with the standard alloy. The tensile strength was improved by 65%, the yield strength by 45% and elongation by 14%. The mechanism by which we achieved the better mechanical and corrosion properties is explained.

Keywords: aluminum alloy AA5083; rapid solidification; melt spinning; high-strength aluminum; extrusion; bimodal microstructure; precipitations

Citation: Paulin, I.; Donik, Č.; Cvahte, P.; Godec, M. Bimodal Microstructure Obtained by Rapid Solidification to Improve the Mechanical and Corrosion Properties of Aluminum Alloys at Elevated Temperature. *Metals* **2021**, *11*, 230. https://doi.org/10.3390/met11020230

Academic Editor: Franc Zupanič
Received: 22 December 2020
Accepted: 26 January 2021
Published: 29 January 2021

Publisher's Note: MDPI stays neutral with regard to jurisdictional claims in published maps and institutional affiliations.

1. Introduction

Aluminum alloys are increasingly popular in the transport industry because of their good corrosion resistance, very good strength-to-weight ratio, and the consequent reduced CO_2 footprint. The most often used alloys with the best mechanical properties are the 2xxx and 7xxx series, which possess high strength and ductility, which benefits from the precipitation strengthening, but have a relatively poor corrosion resistance. In contrast, alloys from the 5xxx series have a good corrosion resistance, but only moderate strength.

High-strength aluminum alloys that are stable at elevated temperatures (above 200 °C) cannot be produced commercially using conventional routes, if their high strength properties are to be based only on precipitation strengthening. It is common that a distinctive loss of mechanical properties is observed at moderate temperatures (above 150 °C) due to over-aging effects. Aluminum alloys that are resistant to high temperature must have a stable microstructure, which is achieved by having stable, incoherent particles in the matrix. These represent barriers to the sliding of dislocations and can be introduced by a rapid solidification of the melt, where smaller insoluble particles remain in the matrix. These particles prevent the movement of the sliding planes and the moving/migration of dislocations. To improve the thermal stability of aluminum alloys, low-diffused transition metals (TMs) such as Fe, Mn, Cr, Ni, V, Co, and/or Mo must be introduced.

Previous studies confirmed the benefits of TM elements for high-strength aluminum alloys at elevated temperature through the addition of specific TM elements, either individually [1] or in pairs [2]. Some studies were performed with fixed amounts of one or two TM elements and the variation of a third element [3]. All the studies used pure aluminum with a specific addition of TM elements. Very few authors [2,4] highlight the importance of using scrap aluminum with contamination in the form of elements mixtures, which have a role in improving thermal stability [5–7]. The diffusivity and solubility of such elements in aluminum is very low, and during conventional production routes, hard and brittle particles of Al-TM intermetallic phases are formed. These large phases reduce the mechanical properties [8,9]. The solution is to dissolve the elements in an aluminum matrix or disperse them in fine intermetallic particles, which can be produced by increasing the solidification rate through melt spinning (cooling rate 10^4–10^6 K/s) or atomization (cooling rate 10^2–10^4 K/s) [10]. Rapid solidification (RS) leads to interesting features, like a significant increase of the alloying elements' solid solubility, a greater refinement of the microstructural grains and the formation of a variety of non-equilibrium aperiodic phases. [11] The equilibrium solubility of TM elements in an aluminum lattice is very low, commonly around 0.03 at. %, and consequently they cannot bring about any effective strengthening with a conventional thermal treatment. The microstructure of rapidly solidified alloys consists of supersaturated solid solution of alloying elements in aluminum, with stable, metastable and quasicrystalline intermetallic phases. [3,12]

The previous investigations and trials attempting to improve the mechanical properties at elevated temperatures for aluminum alloys tended to use pure alloy elements and studied and improved only some of the properties, mostly mechanical, but did not produce materials that would improve a range properties, including corrosion, at the same time. Although it is known that, normally, a major improvement in the mechanical properties leads to moderate corrosion properties [13], and vice versa. Excessive amounts of TM elements have also been used to realize a large increase in the mechanical properties. But the explanation of the strengthening mechanism did not make it clear that, for example, with an excessive amount of added TM elements, despite the high cooling rate, larger Al-TM phases can occur, which negatively affect the corrosion rate.

The main objective of our study was to importantly improve the mechanical and corrosion properties of the aluminum alloy AA5083, which is known as a good corrosion-resisting material with moderate strength. With minor alloy modifications and a rapid solidification route we intended to modify the microstructure and consequently improve the mechanical and corrosion properties at elevated temperature which would classified these materials as high-strength aluminum alloys.

2. Materials and Methods

The commercial aluminum alloy AA5083 (EN AW 5083) was modified with the alloying elements, rapidly solidified, and then isostatically pressed at room temperature, with a final hot extrusion (Figure 1).

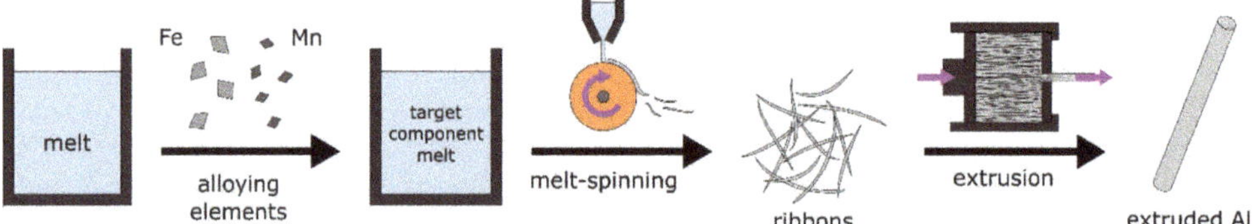

Figure 1. Scheme of the material's preparation.

The chemical compositions of the standard AA5083 aluminum alloy and the modified samples with the addition of only Mn as well as Fe and Mn are presented in Table 1. The most important feature is to apply the modification of the elements up to the allowed tolerance limit according to standard. This is important for industry, because, for example, the transport industry is extremely strict regarding safety standards and is not permitted to introduce new materials until there has been at least 10 years of testing. By staying within the limits of the standard chemical composition, the new, improved material can be more readily accepted.

Table 1. Chemical composition of studied materials (in wt. %).

Samples	Si	Mn	Fe	Zn	Mg	Ti	Al
Standard AA5083 [14]	max. 0.40	0.40–1.00	max. 0.40	max. 0.25	4.00–4.90	max. 0.15	balance
Master AA5083 alloy	0.20	0.52	0.34	0.024	4.3	0.016	balance
RS AA5083	0.20	0.52	0.32	0.024	4.3	0.016	balance
RS AA5083 with Mn	0.20	1.00	0.31	0.024	4.3	0.016	balance
RS AA5083 with Fe and Mn	0.20	1.00	0.45	0.024	4.3	0.016	balance

Melt spinning was used to prepare rapidly solidified ribbons. The aluminum melt was superheated to 1400 °C and cast through small orifice onto a water-cooled copper wheel rotating with a circumferential speed of 50 m/s.

The obtained ribbons were cold compacted under pressure of 200 MPa into billets of 24 mm diameter and approximately 70 mm height. The billets were preheated at the temperature of 420 °C for 15 min prior to the extrusion. The reduction was from 24 mm to 6 mm and the reduction ratio was 16. For comparison, an industrially-extruded standard AA5083 alloy (our master alloy) was used.

The microstructure of the obtained ribbons was first characterized by light microscopy (LM). The samples were prepared by standard metallographic procedure by grinding and polishing with 1-μm surface finishes. To reveal the microstructure for LM the samples were etched using the Weck two-stage etching process. The LM investigation was performed with a Microphot FXA (Nikon Instruments Inc., Tokyo, Japan), an Olympus DP73 camera (Olympus Europe Holding GMBH, Hamburg, Germany) and the Stream Motion (Olympus, Tokyo, Japan) computer program. The microstructure investigations of the ribbons and the extruded materials were performed by scanning electron microscopy (SEM, JEOL FESEM JSM 6500F, (JEOL, Tokyo, Japan)) using electron back-scatter diffraction (EBSD, camera HKL Nordlys II with Channel 5 software (Oxford Instruments HKL, Hobro, Denmark)) for the grain size and the orientation, and energy-dispersive spectroscopy (EDS, INCA ENERGY 400, Oxford Instruments, High Wycombe, UK) for the micro-elemental analysis. Samples of the ribbons were prepared with an ion slicer that was perpendicular to the ribbon. The extruded samples were prepared perpendicular to the extrusion direction by grinding and mechanically polishing with 2 min oxide polishing suspension (OPS). The microstructures of selected ribbons were characterized by high-resolution transmission electron microscopy (TEM, JEOL JEM-2100, JEOL, Tokyo, Japan) in bright field (BF) mode coupled with an EDS analyzer (INCA, Oxford Instruments, High Wycombe, UK).

The Vickers microhardness was measured with a TUKON 2100B (Instron, Norwood, MA, US) instrument. For testing of the ribbons, a 0.025 kgf load was used, and for the extruded material a 0.5 kgf load was used. The microhardness measurements were performed on every sample 5 times and the average values are presented.

The chemical analyses were performed using optical emission spectrometry with an inductive coupled plasma (ICP-OES Agilent 720, (Agilent, Santa Clara, CA, USA)).

The corrosion testing of the extruded samples was performed using electrochemical impedance spectroscopy. An exposed 1 cm^2 of extruded Al sample was prepared for the electrochemical testing in 3.5% NaCl solutions. All the samples were mechanically prepared with SiC paper up to 2400 grit and polished to a mirror finish. Before the electrochemical measurements, the samples were washed and rinsed with acetone and washed in deionized water and dried in air. All the solutions for the experiment were made using MERCK (Merck KGaA, Darmstadt, Germany) chemicals and deionized water. A three-electrode system was used for the measurements, an aluminum electrode (working electrode), a saturated calomel electrode (SCE) as the reference electrode and a

Pt mesh as the counter electrode. All the electrochemical measurements were performed using a Potentiostat/Galvanostat BioLogic SP 300 (Biologic, Seyssinet-Pariset, France) with EC-Lab V11.27 software (Biologic Science Instruments, Seyssinet-Pariset, France). The measurements of all the potentiodynamic polarizations were conducted at a scan rate of 1 mV/s. The data were gathered for the determination of the electrochemical parameters: Corrosion current density i_{corr} and corrosion potential E_{corr} to make the comparison of the extruded samples.

Mechanical tests were performed on an INSTRON 1255 machine (500 kN, Instron, High Wycombe, Buckinghamshire, UK) at room temperature with a constant velocity corresponding to an initial strain rate 10^{-2} s^{-1}. For each group of samples, at least five samples were tested. Based on the results the tensile strength (R_m), yield strength ($R_{p0.2}$) and elongation (A) were recorded. The results are presented as average values.

3. Results and Discussions

To achieve the high strength and corrosion resistance of the aluminum alloy, the melt is rapidly solidified in the form of ribbons and further plastically consolidated using extrusion at elevated temperature. Different chemical compositions of the alloys were prepared and melt spun. The products of melt spinning are typically ribbons, approximately 3 mm wide and 50–120 μm thick, as shown in Figure 2a. The melt-spun ribbons were characterized in the longitudinal direction. An etched LM image reveals the microstructure of the RS aluminum alloy (Figure 2b). Columnar grains appeared on the side where the melt touched the copper wheel. The ribbon's surface on this side is flat and the precipitates are very small and not directly visible in LM micrographs. They are seen as etched pits due to the corrosion attack of the small precipitates. The ribbon's surface on the air side is not flat, but rougher and in cross-section shows a wave-like appearance. On this side of the ribbons the grains are smaller and polygonal with larger precipitates, seen as dark dots. The wheel side of the ribbons has a faster cooling rate, which means the solidified melt is supersaturated. The TM alloying elements like Fe and Mn remain in a solid solution and the precipitates are much smaller. The grains are columnar and the growth direction is defined by the direction of the heat gradient, which is perpendicular to the copper wheel. The very fast cooling rate causes no precipitation during the crystal growth and this is the reason why the grains are larger. While the air-side ribbons solidify more slowly and the precipitates have more time to grow. This part of the ribbons is at a certain point still liquid and has many precipitates, which are seeds for the crystallization. That causes the formation of many small grains that begin to grow at almost the same time.

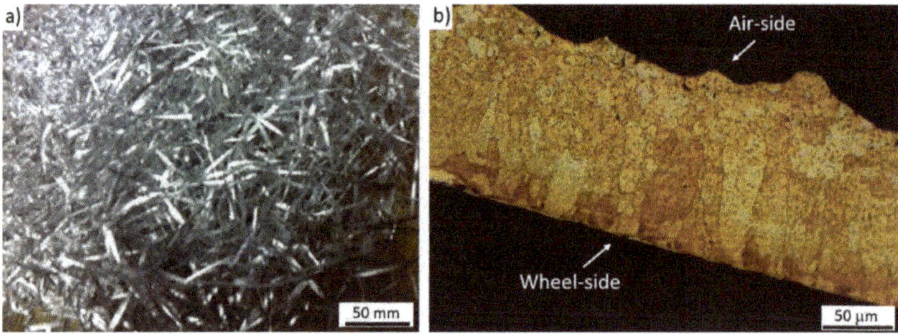

Figure 2. (**a**) Melt-spun ribbons, (**b**) longitudinal cross-section of ribbon (light microscopy (LM), etched by two-stage Weck).

SEM images reveal the microstructure of ribbons even better. Depending on the chemical composition, some ribbons have less or no precipitates, as shown in Figure 3a, where there is a cross-section of the ribbon with no addition of Fe and/or Mn. The addition of TM elements, Mn in Figure 3b and Mn and Fe in Figure 3c, causes more precipitation. As already described for Figure 2b, the majority of ribbons have larger precipitates on the air side; however, it sometimes happens that the ribbons touch each other or the time of the contact with copper wheel can be different. Due to very uneven conditions during melt spinning, the large precipitates are not always on the air side.

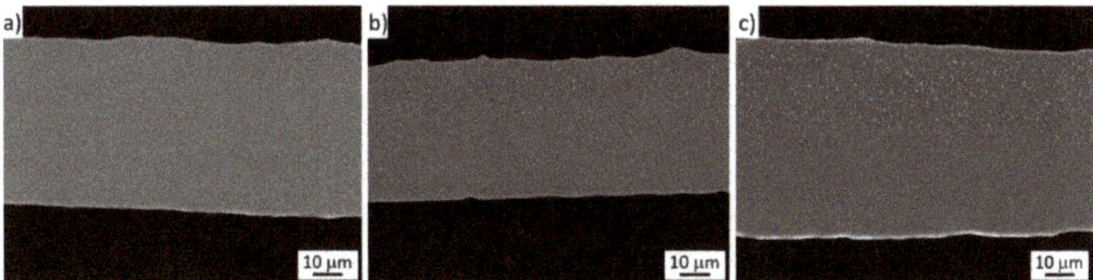

Figure 3. SE images of the cross-section of the melt-spun ribbons (**a**) rapid solidification (RS) AA 5083, (**b**) RS AA 5083 with Mn, (**c**) RS AA5083 with Mn and Fe.

Figure 4 shows the air-side ribbon with bright and dark precipitates analyzed by SEM/EDS. In SE image, the areas of the EDS analyses are marked and the related chemical composition are presented in the inset table. Even though EDS analysis are rather inaccurate for small phases we assume that fine bright phases, according to our EDS analysis and the literature [3,15], are some of these possible phases; α-Al$_{15}$(FeMn)$_3$Si, β-Al$_5$FeSi, Al$_9$Mn$_3$Si and/or α-Al$_{12}$Fe$_3$Si. There are many of these bright colored phases in the microstructure. There is less dark phase (Mg$_2$Si), which is present only in the RS ribbons that were the most slowly cooled. All the phase sizes are submicron, mostly smaller than 500 nm. There are no large phases in the RS ribbons. In the inset table of EDS results in Figure 4 highlighted elements shows the possible bright (Spectra 1 and 2) and dark (Spectra 3 and 4) phases.

EDS results in wt.%	Mg	Al	Si	Mn	Fe	Cu
Spectrum 1	4.6	82.3	0.6	**2.9**	**8.6**	1.0
Spectrum 2	4.2	87.3	0.7	**2.7**	**4.4**	0.7
Spectrum 3	**7.8**	87.6	**3.2**	0.6	0.3	0.5
Spectrum 4	**6.2**	91.0	**2.0**	0.5	0.0	0.3
Spectrum 5	3.5	95.8	0.0	0.3	0.0	0.4

Figure 4. SE micrograph with marked EDS analyses and corresponding EDS results.

TEM BF micrographs were recorded for all three studied samples and both wheel-side and air-side areas. In all the studied areas small precipitates were present and the amount of those precipitates increased with the addition of TM elements. TEM/EDS analyses were performed and in Figure 5 is a TEM BF micrograph with a lot of nanosized precipitates of the sample RS AA5083 with Mn and Fe. In the TEM BF micrographs, large amounts of precipitates based on AlMnFe (Al_6(Mn,Fe) phase) are present in a spherical shape of 50 nm and needles of 50 nm thick and approximately 100–500 nm long. Those precipitates are formed because of the supersaturated TM elements in the rapidly solidified aluminum matrix. They were also studied by Stan-Glowinska et al. [3].

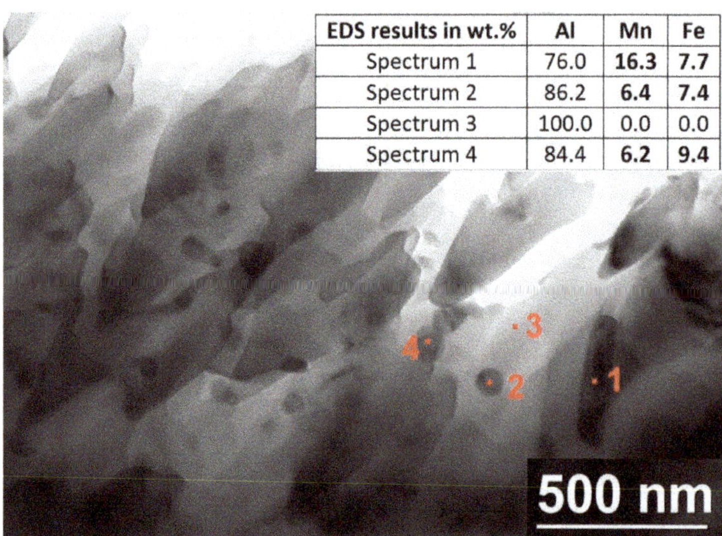

EDS results in wt.%	Al	Mn	Fe
Spectrum 1	76.0	**16.3**	**7.7**
Spectrum 2	86.2	**6.4**	**7.4**
Spectrum 3	100.0	0.0	0.0
Spectrum 4	84.4	**6.2**	**9.4**

Figure 5. TEM bright field (BF) micrograph with marked analyzed EDS areas and corresponding EDS results.

The grain morphologies of the as-melt-spun and the annealed ribbons are clearly seen from the EBSD inverse pole figure in the Z direction (IPF-Z) map (Figure 6), similar to the findings of Tewari [16] using different alloy. The as-melt-spun ribbons in contact with the rotating wheel have larger and columnar grains. The other part solidified subsequently and has a much finer, polygonal grain structure (Figure 6a). Because the solidification starts at a large number of seeds at the same time, a very fine grain structure appears, with polygonal shapes. Due to the different solidification conditions of the ribbons, the bimodal structure is obtained. In the literature [2], quite often the formation of fine grains is correlated with the wheel side, because this part of the ribbons is much more quickly cooled and solidified, which seems to be contradictory. The very flat surface of the ribbon is undoubtedly proof of wheel side, which is correlated with columnar grains.

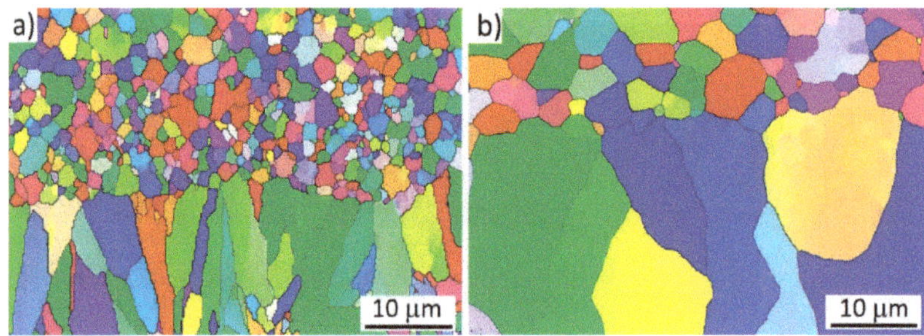

Figure 6. SEM electron back-scatter diffraction (EBSD) IPF-Z maps of cross-section of melt spun ribbons with addition of Mn and Fe, (**a**) as melt spun, (**b**) after annealing at 500 °C for 3 h.

The temperature stability of the ribbons was tested by exposure to different temperatures (T interval from 350 to 500 °C) and times (t = 3, 12, 24 h) and was performed to select the proper conditions for extrusion. No significant differences by measuring the microhardness were revealed up to 450 °C. The first drop in the microhardness of the ribbons was detected at 470 °C. The microstructure (Figure 6b) of the annealed ribbons at 500 °C for 3 h with the addition of Mn and Fe shows a slight difference in the grain shape, because they started to coarsen due to the high temperature and longer exposure time.

A comparison of the three RS ribbons with modified chemical compositions shows that the most temperature-resistant material is the one with the addition of Mn and Fe. Microhardness measurements of those ribbons show the highest values for all the temperature-testing ranges (Table 2).

Table 2. Vickers microhardness of ribbons (HV0.025) for each annealing temperature with three different annealing times.

Annealing T (°C)	350	400	450	470	500	as-spun
Annealing t (h)	3/12/24	3/12/24	3/12/24	3/12/24	3/12/24	3/12/24
RS AA5083	96/87/87	96/83/82	89/84/81	82/81/81	79/81/75	98
RS AA5083 + Mn	97/97/96	98/95/96	95/94/92	87/83/84	82/79/78	105
RS AA5083 + Mn + Fe	107/104/104	101/102/101	98/96/95	89/89/87	87/85/83	109

A bimodal microstructure is achieved by rapid solidification due to the nature of the melt spinning's solidification and consists of two different regions: The wheel side and the air side. The wheel side is cooled faster and is composed of a supersaturated solid solution of TM elements and has larger columnar grains. On the other hand, the air side is less intensively cooled and consists of stable and metastable phases and has smaller grains, mainly under 1 μm. Voderova et al. [2,10] discuss the phases and precipitations achieved by melt spinning of aluminum alloys with addition of different TM elements. The rapidly solidified aluminum alloy with the addition of Fe (up to 11 wt. %) is composed of a supersaturated solid solution of Fe in aluminum and metastable Al_6Fe [17] on the faster-cooling side, while the air side contains a smaller amount of stable $Al_{13}Fe_4$, also referred as Al_3Fe [18], and metastable Al_6Fe. The supersaturated TM elements in our study were confirmed by TEM analyses in Figure 5, where all the nano-precipitates consist of Al, Mn and Fe. On the other hand, the air side of the ribbons contained $Al_{13}(Mn,Fe)_4$, $Al_{15}(Mn,Fe)_3Si$ and Mg_2Si phases, which were analyzed by SEM/EDS in Figure 4.

After the melt-spun material was cold compacted into billets the hot extrusion was performed at 420 °C. Cross-sections perpendicular to the extrusion direction were prepared for EBSD microstructure analyses. Figure 7 shows the microstructure development during the extrusion of the RS ribbons. In Figure 7a is extruded RS AA5083 material with a homogeneous grain size and distribution. The grains of RS AA5083 with Mn (Figure 7b) and RS AA5083 with Mn and Fe (Figure 7c) show a bimodal gran structure where both larger and smaller grains in ribbons, despite hot extrusion, retain their size and shape. The effect of the bimodal structure appears due to the very good thermal stability of the RS material with the addition of TM elements. Nanoprecipitates based on TM elements keep the material stable by preventing the grain refinement of larger grains due to the mechanical deformation and grain growth due to the recrystallization temperature. The TM elements prevent the migration of dislocations and make it difficult for the sliding planes to move.

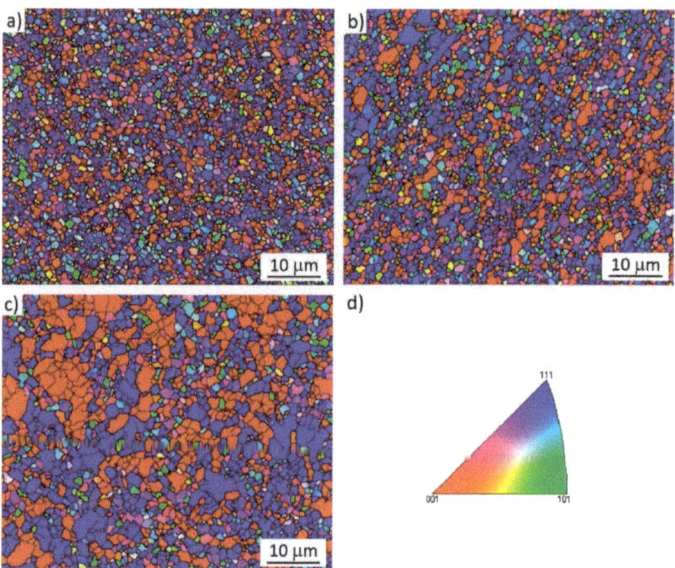

Figure 7. EBSD IPF Z maps of cross-section of the extruded material (**a**) IPF-Z of RS AA5083, (**b**) RS AA5083 with Mn, (**c**) IPF-Z of RS AA5083 with Mn and Fe, and (**d**) IPF legend.

The mechanical tests of the RS extruded material showed a significant improvement in the tensile and yield strength without decreasing the elongation. In comparison to the master alloy (industrially extruded) AA5083 material and the standard tensile strength of the extruded RS AA5083 increase for 5% and 30%, respectively, and for extruded RS AA5083 material with Mn and with Mn and Fe the increases were 20% and 48%, respectively, for Mn addition and 34% and 65% for Mn and Fe addition. Also, the yield strength of the extruded RS AA5083 with Mn and extruded RS AA5083 with Mn and Fe increased by 30% and 125% for the Mn addition and 45% and 151% for Mn and Fe additions, respectively. The elongation of the extruded RS AA5083 with Mn and Fe increased by 100% compared to the standard material and 14% compared to industrial extruded AA5083 master alloy. The average results of the mechanical tests are presented in Table 3.

Table 3. Mechanical properties of standard AA5083 compared with the studied materials.

Samples	Tensile Strength R_m (MPa)	Elongation A (%)	Yield Strength $R_{p0.2}$ (MPa)	HV0.5
Standard for AA5083 [14]	270	12	125	75
Extruded AA5083	334	20	216	90
Ex. RS AA5083	353	21	195	97
Ex. RS AA5083 with Mn	402	20	282	110
Ex. RS AA5083 with Mn and Fe	447	24	314	112

The potentiodynamic measurements of the four different extruded Al samples in a typical corrosion medium of 3.5% NaCl is shown in Figure 8 and Table 4. The corrosion rate calculations were performed according to the Faraday's law described in ASTM G102-89 (re-approved in 2015 standard) [19]:

$$v_{\text{corr}} = K_1 \left(i_{\text{corr}} / \rho \right) EW \tag{1}$$

where K_1 is 3.27 10^{-3} mm/g μA cm yr (for mmpy units of v_{corr}), i_{corr} is corrosion current density in μA/cm^2, which is calculated from the Stern–Geary equation [19] by B/R_p, where B is calculated and R_p is the polarization resistance from the slope of the potential versus current density plot taken approximately 20 mV on either side of the open circuit potential, ρ is density of the material in g/cm^3 and EW is equivalent weight (considered dimensionless in this equation), while i_{corr} is in ohm-cm^2. Rp is the slope of the potential versus current density plot taken approximately 20 mV on either side of the open circuit potential where the slope is often approximately linear.

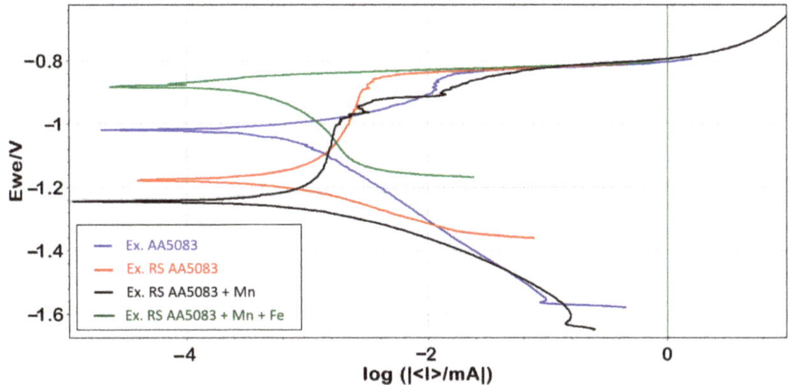

Figure 8. Polarization curves recorded for four different extruded RS Al samples in 3.5% NaCl.

Table 4. Corrosion properties of the studied extruded materials.

Samples	E_{corr} vs.SCE (V)	i_{corr} (μA)	v_{corr} (mm/y)	Corrosion Rate (mm/y)
Ex. AA5083	−1.019	0.98	3.78×10^{-2}	0.0378
Ex. RS AA5083	−1.176	1.11	4.28×10^{-2}	0.0428
Ex. RS AA5083 with Mn	−1.251	0.64	2.48×10^{-2}	0.0248
Ex. RS AA5083 with Mn and Fe	−0.857	0.16	6.21×10^{-3}	0.0062

The main difference between the samples is the chemical composition and the rate of solidification (normal material extrusion and extrusion of RS material) of the alloys, which all affected the polarization potential and the passivation behavior. Before all the measurements of the corrosion parameters, 1 h of stabilization at the open-circuit potential (OCP) occurred. Corrosion potentials (E_{corr}) for these samples changed quite fairly, as the RS AA5083 with Mn has E_{corr} around −1.2 V_{SCE} and RS AA5083 with Mn and Fe sample has approximately −0.86 V_{SCE}. On the other hand, following the Tafel region, the alloys exhibited a differently broad range of passivation with practically the same breakdown potential (E_b) at around −0.83 V_{SCE}. The passivation range is significantly narrowed, and there is almost no passive region for sample RS AA5083 with Mn and Fe, whereas the corrosion-current density (i_{corr}) and corrosion rate (v_{corr}) was almost seven-times lower compared to the highest i_{corr} and v_{corr} of the sample RS AA5083. In the passive range the corrosion-current densities changed for the tested specimens from 0.16 mA to almost 1.11 mA. In the active-passive transition just two of the studied specimens went into the typical passive range, RS AA5083 and RS AA5083 with Mn, whereas the other two did not exhibit a typical passive region or it was considerably narrower.

The correlation between grain size and corrosion in the Al alloys has been extensively studied [20–23], but the literature data does not present a unified explanation for the effect of grain refinement on the corrosion resistance of Al alloys. On the other hand, the amount of intermetallic inclusions usually plays a crucial role in the susceptibility of different Al alloys to corrosion characteristics [24,25]. However, the effects of grain boundaries and intermetallic phases on the corrosion are still not thoroughly understood and each play an important role in the final corrosion characteristics of the alloy.

The corrosion-favorable parameters in our studied Al alloys correlate with the grain size and consequently with the number/length of the grain boundaries where the inclusions less favorable to corrosion can form (e.g., β-phase (Al_3Mg_2)) with an increased amount of Mg that migrates to the grain boundaries [25]. In our case, the corrosion rate increases with the number of grains. As a result, our studied alloys corroborate with the data described for the similarly processed Al alloys (severe plastic deformation SVD, high-pressure torsion HPT, equal-channel angular pressing ECAP, friction steer welding FSW, etc.) [22,26]. The improved corrosion resistance in our study occurs in the samples with a bimodal microstructure. The EBSD data (Figure 7) show that the number of grains per analyzed area in the samples with a bimodal microstructure is reduced by approximately 50%. The smaller number of grains in the extruded RS AA5083 samples with the addition of Mn and Fe is due to partly the larger grains in the material's microstructure. The material with addition of Mn and Fe has larger amount of bimodal microstructure (Figure 7c) and also larger amount of grains with (001) planes parallel to the surface. From the literature [27] it is known that (001) single crystals has most noble pitting potential values. This can be explanation why material with Mn and Fe has despite the additional elements even better corrosion resistance.

4. Conclusions

The study was focused on the preparation of a high-strength aluminum alloy with superior mechanical properties and very good corrosion resistance at elevated temperature within the limits of the standard AA5083 alloy. By using the melt-spinning technique we achieved supersaturated, temperature-resistant ribbons from which the extruded aluminum material was produced. With the addition of transition-metal elements to the standard alloy we improved the mechanical and corrosion properties with a bimodal microstructure of the material.

The main achievements and conclusions are:

- The explanation of the grain morphology was related to the nature of the melt-spinning process—the grains are larger and columnar in the direction of the temperature gradient on the wheel side of the RS ribbons. On the other hand, the ribbons' air side consists of a larger amount of smaller grains, typically around 1 µm.
- The explanation of the precipitates was related to the nature of melt-spinning process—the precipitates on the wheel side of the RS ribbons are nano-sized, based on AlMnFe, due to the extremely high cooling rate. The air side of the RS ribbons consists of, besides the nano-sized AlMnFe precipitates, also micron-sized stable and metastable phases, like Mg_2Si, and phases based on AlMnFeSi.
- The relation of the chemical composition to the grain morphology after extrusion—the addition of transition-metal elements generates in the matrix material nano-sized precipitates, which influenced the grain behavior during extrusion. The material with more precipitates retains the bimodal structure due to melt spinning, independent of the subsequent deformation process, even at elevated temperatures. This makes the material stable at higher temperature.
- The relationship between the chemical composition and the mechanical and corrosion properties depends on the addition of transition-metal elements directly and indirectly influences the nano- and microstructure of the RS material, as well as the grain behavior during the thermal treatment. The optimal concentrations of TM elements improve all the mechanical properties and maintain or even improve the corrosion resistant of the RS aluminum alloy.

Based on experiments, we have shown that it is possible to produce aluminum alloys with high-strength properties and excellent corrosion resistance at elevated temperatures, which can also be introduced and up-scaled into an industrial manufacturing process.

Author Contributions: Conceptualization, I.P. and M.G.; methodology, I.P., M.G., and P.C.; software, I.P., and Č.D.; validation, I.P., M.G., and Č.D.; formal analysis, I.P., Č.D., and M.G.; investigation, I.P., M.G., and P.C.; writing—original draft preparation, I.P., M.G., and Č.D.; writing—review and editing, I.P. and M.G.; supervision, M.G. and I.P.; funding acquisition, P.C. and M.G. All authors have read and agreed to the published version of the manuscript.

Funding: This research was funded by the Slovenian Research Agency, research core funding No. P2-0132. This work was also funded by the project program "Optimization of a new generation of high-strength corrosion-resistant aluminum alloys at elevated temperatures based on rapid solidification technology", supported by IMPOL 2000 d.d. aluminum industry.

Data Availability Statement: The data presented in this study are available on request from the corresponding author.

Conflicts of Interest: The authors declare no conflict of interest.

References

1. Cavojsky, M.; Svec, P., Sr.; Janickovic, D.; Orovcik, L.; Simancik, F. Rapidly solidified Al-Mo and Al-Mn ribbons: Microstructure and mechanical properties of extruded profiles. *Kov. Mater* **2014**, *52*, 371–376. [CrossRef]
2. Voderova, M.; Novak, P.; Marek, I.; Vojtech, D. *Microstructure and Mechanical Properties of Rapidly Solidified Al-Fe-X Alloys. Materials Structure & Micromechanics of Fracture VII*; Sandera, P., Ed.; Trans Tech Publications Ltd.: Pfaffikon, Switzerland, 2014; Volumes 592–593, pp. 639–642.
3. Stan-Głowińska, K.; Lityńska-Dobrzyńska, L.; Morgiel, J.; Góral, A.; Gordillo, M.A.; Wiezorek, J.M. Enhanced thermal stability of a quasicrystalline phase in rapidly solidified Al-Mn-Fe-X alloys. *J. Alloys Compd.* **2017**, *702*, 216–228. [CrossRef]
4. Voderova, M.; Novak, P.; Vojtech, D. Microstructure of rapidly solidified and hot-pressed Al-Fe-X alloys. *Mater. Tehnol.* **2014**, *48*, 349–353.
5. Lojen, G.; Boncina, T.; Zupanic, F. Thermal stability of Al-Mn-Be melt-spun ribbons. *Mater. Tehnol.* **2012**, *46*, 329–333.
6. Školáková, A.; Hanusová, P.; Průša, F.; Salvetr, P.; Novák, P.; Vojtěch, D. Microstructure and thermal stability of Al-Fe-X alloys. *Acta Metall. Slovaca.* **2018**, *24*, 223–228. [CrossRef]
7. Voderova, M.; Novak, P.; Michalcova, A.; Vojtech, D. Thermal stability of rapidly solidified Al-Fe-X alloys. In Proceedings of the METAL 2013: 22nd International Conference on Metallurgy and Materials, Brno, Czech Republic, 15–17 May 2013; pp. 1161–1164.
8. Karakose, E.; Keskin, M. Structural investigations of mechanical properties of Al based rapidly solidified alloys. *Mater. Des.* **2011**, *32*, 4970–4979.
9. Tokarski, T.; Wedrychowicz, M. High strength 5083 aluminium alloy via rapid solidification and plastic consolidation route. In Proceedings of the METAL 2015: 24th International Conference on Metallurgy and Materials, Brno, Czech Republic, 3–5 June 2015; pp. 1215–1220.
10. Voderova, M.; Novak, P.; Prusa, F.; Vojtech, D. Microstructures of the Al-Fe-Cu-X alloys prepared at various solidification rates. *Mater. Tehnol.* **2014**, *48*, 771–775.
11. Inoue, A.; Kong, F.L.; Zhu, S.L.; Shalaan, E.; Al-Marzouki, F.M. Production methods and properties of engineering glassy alloys and composites. *Intermetallics* **2015**, *58*, 20–30.
12. Galano, M.; Audebert, F.; Garcia Escorial, A.; Stone, I.C.; Cantor, B. Nanoquasicrystalline Al-Fe-Cr-based alloys with high strength at elevated temperature. *J. Alloys Compd.* **2010**, *495*, 372–376. [CrossRef]
13. Halap, A.; Popovic, M.; Radetic, T.; Vascic, V.; Romhanji, E. Influence of the thermos-mechanical treatment on the exfoliation and pitting corrosion of an AA5083-type alloy. *Mater. Tehnol.* **2014**, *48*, 479–483.
14. The Aluminum Association. *International Alloy Designations and Chemical Composition Limits for Wrought Aluminum and Wrought Aluminum Alloys*; The Aluminum Association: Arlington, VA, USA, 2015.
15. Li, Z.; Argyropoulos, S.A. The Assimilation Mechanism of Mn-Al Compacts in Liquid Mg. *Mater. Trans.* **2010**, *51*, 1371–1380. [CrossRef]
16. Tewari, S.N. Effect of melt spinning on grain-size and texture in Ni-Mo alloys. Metall. Trans. A-PHYSICAL Metall. *Mater. Sci.* **1988**, *19*, 1711–1720.
17. Michalcová, A.; Marek, I.; Knaislová, A.; Veselka, Z.; Vavřík, J.; Bastl, T.; Hrdlička, T.; Kučera, D.; Vojtěch, D. Aluminium Based Alloys with High Content of Metastable Phases Prepared by Powder Metallurgy. *ACTA Phys. Pol. A* **2018**, *134*, 688–691. [CrossRef]
18. Chen, C.; Wang, J.; Shu, D.; Li, P.; Xue, J.; Sun, B. A Novel Method to Remove Iron Impurity from Aluminum. *Mater. Trans.* **2011**, *52*, 1629–1633. [CrossRef]
19. ASTM G102-89(2015)e1. *Standard Practice for Calculation of Corrosion Rates and Related Information from Electrochemical Measurements*; ASTM International: West Conshohocken, PA, USA, 2015; Available online: www.astm.org (accessed on 12 November 2020).

20. Ralston, K.D.; Birbilis, N. Effect of Grain Size on Corrosion: A Review. *Corrosion* **2010**, *66*, 075005. [CrossRef]
21. Kus, E.; Lee, Z.; Nutt, S.; Mansfeld, F. A comparison of the corrosion behavior of nanocrystalline and conventional Al 5083 samples. *Corrosion* **2006**, *62*, 152–161. [CrossRef]
22. Nakano, H.; Yamaguchi, H.; Yamada, Y.; Oue, S.; Son, I.J.; Horita, Z.; Koga, H. Effects of High-Pressure Torsion on the Pitting Corrosion Resistance of Aluminum-Iron Alloys. *Mater. Trans.* **2013**, *54*, 1642–1649. [CrossRef]
23. Beura, V.K.; Kale, C.; Srinivasan, S.; Williams, C.L.; Solanki, K.N. Corrosion behavior of a dynamically deformed Al-Mg alloy. *Electrochim. Acta* **2020**, *354*, 136695. [CrossRef]
24. Yasakau, K.A.; Zheludkevich, M.L.; Lamaka, S.V.; Ferreira, M.G.S. Role of intermetallic phases in localized corrosion of AA5083. *Electrochim. Acta* **2007**, *52*, 7651–7659. [CrossRef]
25. Zhu, Y.; Sun, K.; Frankel, G.S. Intermetallic Phases in Aluminum Alloys and Their Roles in Localized Corrosion. *J. Electrochem. Soc.* **2018**, *165*, C807–C820.
26. Jain, S.; Hudson, J.L.; Scully, J.R. Effects of constituent particles and sensitization on surface spreading of intergranular corrosion on a sensitized AA5083 alloy. *Electrochim. Acta* **2013**, *108*, 253–264. [CrossRef]
27. Treacy, G.M.; Breslin, C.B. Electrochemical studies on single-crystal aluminium surfaces. *Electrochim. Acta* **1998**, *43*, 1715–1720. [CrossRef]

Article

Role of Small Addition of Sc and Zr in Clustering and Precipitation Phenomena Induced in AA7075

Martin Vlach [1,*], Veronika Kodetova [1], Jakub Cizek [1], Michal Leibner [1], Tomáš Kekule [1], František Lukáč [1,2], Miroslav Cieslar [1], Lucia Bajtošová [1], Hana Kudrnová [1], Vladimir Sima [1], Sebastian Zikmund [1], Eva Cernoskova [3], Petr Kutalek [3], Volkmar-Dirk Neubert [4] and Volkmar Neubert [4]

[1] Faculty of Mathematics and Physics, Charles University, Ke Karlovu 3, CZ-12116 Prague, Czech Republic; veronika.kodetova@seznam.cz (V.K.); jakub.cizek@mff.cuni.cz (J.C.); mleibner@seznam.cz (M.L.); tomas.kekule@mff.cuni.cz (T.K.); lukac@ipp.cas.cz (F.L.); cieslar@met.mff.cuni.cz (M.C.); lucibajtos@gmail.com (L.B.); hana.kudrnova@mff.cuni.cz (H.K.); vladimir.sima@mff.cuni.cz (V.S.); sebastienz@seznam.cz (S.Z.)

[2] Institute of Plasma Physics, The Czech Academy of Sciences, Za Slovankou 3, CZ-18200 Prague, Czech Republic

[3] Joint Laboratory of Solid State Chemistry, Faculty of Chemical Technology, University of Pardubice, Studentská 84, CZ-53210 Pardubice, Czech Republic; eva.cernoskova@upce.cz (E.C.); petr.kutalek@upce.cz (P.K.)

[4] Institute for Materials Testing and Materials Engineering, Freiberger Strasse 1, D-38678 Clausthal-Zellerfeld, Germany; info@mse-clausthal.de (V.-D.N.); volkmar.neubert@tu-clausthal.de (V.N.)

* Correspondence: martin.vlach@mff.cuni.cz; Tel.: +42-022-191-1459

Abstract: A detailed characterization of phase transformations in the heat-treated commercial 7075 aluminum alloys without/with low Sc–Zr addition was carried out. Mechanical and electrical properties, thermal and corrosion behavior were compared to the microstructure development. The eutectic phase consists of four parts: $MgZn_2$ phase, Al_2CuMg phase (*S*-phase), $Al_2Zn_3Mg_3$ phase (*T*-phase), and primary λ-Al(Mn,Fe,Si) phase. Strengthening during non-isothermal (isochronal) annealing is caused by a combination of formation of the GP zones, η'-phase, *T*-phase and co-presence of the primary and secondary Al_3(Sc,Zr)-phase particles. Positive influence on corrosion properties is owing to the addition of Sc–Zr. Positron annihilation showed an evolution of solute Zn,Mg-(co-)clusters into (precursors of) the GP zones in the course of natural ageing. The concentration of the (co-)clusters is slightly negatively affected by the low Sc–Zr addition. A combination of both precipitation sequences of the Al–Zn–Mg–Cu-based system was observed. The apparent activation energy values for dissolution/formation of the clusters/GP zones and for formation of the metastable η'-phase, stable *T*-phase and stable η-phase were calculated.

Keywords: 7xxx aluminum series; early precipitation stages; corrosion; Al_3(Sc,Zr) particles; annihilation of positrons; activation energy

Citation: Vlach, M.; Kodetova, V.; Cizek, J.; Leibner, M.; Kekule, T.; Lukáč, F.; Cieslar, M.; Bajtošová, L.; Kudrnová, H.; Sima, V.; Zikmund, S.; et al. Role of Small Addition of Sc and Zr in Clustering and Precipitation Phenomena Induced in AA7075. *Metals* **2021**, *11*, 8. https://dx.doi.org/10.3390/met11010008

Received: 25 November 2020
Accepted: 21 December 2020
Published: 23 December 2020

Publisher's Note: MDPI stays neutral with regard to jurisdictional claims in published maps and institutional affiliations.

1. Introduction

The heat-treated commercial aluminum 7xxx series (Al–Zn–Mg–Cu-based) alloys extensively appear in various applications, including metalworking, automotive, aircraft, and space engineering thanks to their precipitation-hardening reaction and light weight [1–5]. Since aluminum (Al) is commonly used nonferrous metal, the world's consumption of Al increased year by year, mainly driven by the growth in China [2,3]. According to the European Aluminum Association (EAA), massive usage of Al would decrease the CO_2; decreasing the weight of the (hybrid) vehicles by ~100 kg reduces the fuel (energy) consumption ~5 percent [2,3]. Even though aluminum commercial 7xxx-based alloys have been traditionally used especially as a lightweight material for traffic systems, the trends in industry emphasize the improvement and stabilization of the structure and properties of these materials through a careful control of thermo-mechanical processing schedules [4,5].

The chemical composition of these alloys plays a really big role in their characterization. Many of these useful properties can be influenced and controlled by the appropriate alloying of Mg and Zn [6–11]. However, they are regulated not only by the concentration of solutes, but also by processing and suitable parameters including heat treatment/annealing procedure, rolling/deformation, quenching and/or subsequent ageing [7,10–17]. It enables study to vary the concentration of different types of elements dissolved in Al as well as the size distribution/composition of phases.

The category of transformation line is changed with the content and composition of a particular element; the (non-)isothermal series of annealing steps are usually used [6,8,10,17–23]. Despite having some disagreements and obscurity about the arrangement and the types of the hardening phases in the alloys, the decomposition of supersaturated solid solution is in the consistency with two precipitation sequences, where η'(MgZn$_2$)-phase (hexagonal)/η(MgZn$_2$)-phase (hexagonal) and T'(Al$_2$Zn$_3$Mg$_3$)-phase (hexagonal)/T(Al$_2$Zn$_3$Mg$_3$)-phase (cubic) occur [15,19,20]. The precipitation of the semi-coherent η'- and T'-phases usually cause the age-hardening of the materials in contrast to the incoherent η- and T-phases [19,20,22–27]. Many authors use different designations for the early precipitation phases formed during the precipitation; (co-)clusters/Guinier Preston (GP) zones of different types usually appear during ageing at temperatures lower than ~150 °C [19,20,25,28–33].

One possibility how to improve the potential of traditional commercial aluminum 7075-(Al–Zn–Mg–Cu–Mn–Si–Fe-) based alloys is the addition of rare earth (RE) elements (incl. Sc) and/or the addition of selected transition (TM) metals (e.g., Zr) [1,12–16,19,34–39]. The TM elements diffuse much slower than Sc and a TM-enriched shell behaves as a diffusion barrier preventing the diffusion from the Sc-rich core and this way reduces a coarsening of the core-shell secondary Al$_3$(Sc,TM)-phase [1]. It enhances the usage of these alloys up to ~400 °C. The addition of Sc (0.05–0.20 at.%) and Zr (0.05–0.50 at.%) is the most effective Sc–Zr combination in aluminum due to the precipitation of the secondary spherical (coherent) core-shell ternary L1$_2$-structured Sc,Zr-containing phase [1,12,13,20,34,36]. The Sc and Zr content can help to refine and to affect the microstructure and the recrystallization temperature and therefore the materials in a non-recrystallized state improve their mechanical properties and resistance to corrosion [1,15,40–42]. It has recently been found in the commercial Al alloys with low Sc–Zr addition that the inter-metallic particles of primary Al$_3$(Sc,Zr) phase have a multi-layered structure with many different shapes (dimension about 1–3 μm) and they appear inside (in the center of) the grains in principle [15,20].

The reason of this work is a detailed characterization of the phase transformation in heat-treated Al–Zn–Mg–Cu–Mn–Si–Fe (–Sc–Zr) alloys and thus contribute to the knowledge of the ongoing processes in the aluminum commercial 7xxx alloys. This work is based on ongoing research, which has partially been recently published in References. [15,19,20,25]. The results of positron annihilation, microhardness and resistivity measurements during the isochronal annealing and subsequent natural ageing, thermal response and corrosion behavior were suitably supplemented with the powder X-ray diffraction and microstructure observations by scanning and transmission electron and atomic force microscopy. The article attempts to enlarge the knowledge about the very early precipitation stages, precipitation mechanism, corrosive characteristics and activation energies for the dissolution/formation of the phases in the Al-based commercial alloys with low Sc–Zr addition. Due to the fact that this is a study of a cast heat-treated ("pseudo-homogenised") aluminum 7075 (-Sc–Zr)-based alloy (i.e., without thermo-mechanical processing such as hot rolling or cold rolling), the research results are relevant to the use of the alloys as shape castings or perhaps more significantly for additive manufacturing although the solidification conditions would be very different.

2. Materials and Methods

Two alloys were studied: Without Sc–Zr addition (labelled as 7075 alloy) and with Sc–Zr addition (labelled as 7075-ScZr alloy). The composition of the alloys is listed in Table 1 (Al is the balance).

Table 1. Chemical composition of the alloys investigated (at.%), Al is the balance.

Alloy	Mg	Zn	Cu	Mn	Si	Fe	Sc	Zr
7075	3.44	2.76	0.79	0.21	0.22	0.18	-	-
7075-ScZr	3.41	2.72	0.77	0.26	0.24	0.23	0.14	0.06

The materials were conventionally prepared by melting of pure Al, Mg, Zn, and Cu (the purity of Al used was more than 99.5%, in case of Mg, Zn, and Cu more than 99%) and commercial-purity master alloys (Al-2Sc and Al-10Zr) in a resistance vacuum furnace (SVÚM, Čelákovice, Czech Republic) at ~ 850 °C. The materials were cast into a cylindrical steel mold with size of ~125 mm × 20 mm × 270 mm. The total weight of the ingots was about 1.5 kg; it was therefore an experimental laboratory produced casting. The materials after casting were stored at room temperature (RT) until the measurements. Specimens for characterization of phase development and age hardening were mainly cut parallel to the longitudinal ingot axis. Sampling from the outer ingot part was excluded. The alloys were then subjected to the high temperature annealing at 470 °C in the furnace with a protective Ar-atmosphere. The annealing time at 470 °C was 60 min (or 240 min). After the annealing, the materials were quenched in water at RT. Unless otherwise stated, the state of the alloys after high temperature annealing at 470 °C/60 min is referred to the (heat-treated) HT state. In this view, therefore, the state of the materials after the annealing at 470 °C can be understood as the as-prepared/initial state. The alloys after high temperature annealing at 470 °C were also subjected to the subsequent natural ageing at RT. The HT + NA state of the alloys means the alloys in the HT state after 3500 h (~146 days) of natural ageing (NA). The HT + NA15000 state of the alloys means the alloys in the HT state after 15,000 h (~625 days) of natural ageing. Summary of the states of the alloys is given in the Table 2.

Table 2. Summary of the heat treatment conditions (types of the annealing) and the marking of the alloy states.

State of the Alloys	Heat Treatment
HT	470 °C/60 min
470 °C/240 min	470 °C/240 min
HT + NA	470 °C/60 min + natural ageing for 3500 h
HT + NA15000	470 °C/60 min + natural ageing for 15,000 h
470 °C/240 min + NA	470 °C/240 min + natural ageing for 3500 h

The alloys in the different states were further characterized after the step-by-step heat treatment to the different temperatures. The isochronal sequent heat treatment with effective speed of 1 °C/min was performed either in the oil bath (up to 240 °C) or in the furnace (above 240 °C). The immediate cooling into liquid N_2 or water came after the annealing step. The materials were kept in liquid N_2 at 78 K after quenching until measurements or other annealing started to avoid possible NA.

The (relative) electrical resistivity measurements were determined in liquid N_2 bath at 78 K by means of the DC four-point method. The influence of parasitic thermo-electromotive force was suppressed by current reversal with a dummy specimen in series. H-shaped specimens machined to dimensions of ~2 mm × 9 mm × 90 mm were used for the resistivity measurements; the length represents the gauge length. Vickers microhardness (HV0.5) testing (specimens machined to dimensions of ~20 mm × 20 mm × 20 mm) was performed in Wolpert Wilson Micro Vickers 401MVD (Wilson Instruments, Canton, MA, USA) at ~5 °C to avoid possible NA. Detailed information about electrical resistivity and microhardness HV0.5 measurements is described in References. [19,25].

Thermal properties were studied using differential scanning calorimetry (DSC) performed in Netzsch DSC 204 F1 Phoenix apparatus (NETZSCH-Gerätebau, Selb, Germany) at heating rates of 1–20 °C/min up to 400 °C. A material of diameter 2–4 mm and of mass

12–18 mg was placed in Al_2O_3 crucibles and measured in a dynamic nitrogen atmosphere (20 mL/min).

Powder X-ray diffraction (XRD) in symmetric Bragg–Brentano geometry of the HT state of the alloys (surface area of ~4 cm^2) was carried out on a vertical powder θ-θ diffractometer D8 Discover (Bruker AXS, Karlsruhe, Germany) using CuKα radiation with a NiKβ filter at 40 kV and 40 mA from 15° to 90° (0.02° per step). The measurement was performed at RT.

Positron annihilation spectroscopy (PAS) was employed for positron lifetime (LT) evolution during natural ageing of the HT state of the 7075 and 7075-ScZr alloys (surface area of ~1 cm^2). PAS studies were done at −50 °C using a ^{22}Na positron source (activity of ~1 MBq) sealed between 5 μm thick Ti foils. One measurement lasted about 2 days. A digital spectrometer with time resolution of 145 ps [43] was employed for LT measurement. Detailed information about LT and XRD measurements is described in References [19,20,44].

Corrosion characteristic were studied by potentiodynamic polarization experiments which were carried out using auto range Wenking M Lab Potentiostat (controlled by a computer, Institut für Materialprüfung und Werkstofftechnik, Clausthal-Zellerfeld, Germany). Fresh surfaces of specimens for corrosion testing were obtained by grinding using 1200 grit SiC papers. Specimen with surface area of ~1 cm^2 was immersed in 0.1 M NaCl aqueous solution for 15 min prior to polarization, by which time a stable potential (OCP) was obtained. Saturated calomel electrode was used as s reference electrode. Polarization was obtained by scanning from 500 mV more negative than OCP at rate of 20 mV/min. The measurements were performed at RT.

The surface topography of the HT state of the alloys (surface area of ~4 cm^2) was gauged employing an Atomic Force Microscopy (AFM) Solver Pro-M (NT-MDT, Moscow, Russia) in the semi-contact mode. Microstructure pictures were recorded with the frequency of 0.7 Hz and the resolution 256 × 256 pixels. The measurements were realized using high-resolution etalon PHA_NC type cantilevers (Au coating, cone angle less than 22°; resonant frequency 140 kHz and force constant 3.5 N/m). Microstructure observations of the materials were observed by disposing transmission electron microscopy (TEM), JEOL JEM-2000FX microscope (JEOL, Tokyo, Japan), and scanning electron microscopy (SEM), MIRA I Schottky FE-SEMH microscope (TESCAN ORSAY HOLDING, Brno-Kohoutovice, Czech Republic). The analysis of precipitates was carried out by energy-dispersive spectroscopy (EDS) using an X-ray BRUKER microanalyzer (Bruker AXS, Karlsruhe, Germany). The specimens for TEM and SEM were annealed by the same procedure (isochronal sequent annealing with effective speed of 1 °C/min) as the specimens for electrical resistivity and microhardness HV0.5 measurements.

3. Results and Discussion

3.1. Heat-Treated (Initial) State of the Alloys

Characterization of the microstructure of the heat-treated (initial) state of the materials is very complex, as it is presented below. The HT alloys contain eutectic phases at grain boundaries; eutectic volume fraction was observed comparable for both alloys—an overview is shown in Figure 1. Figure 2 presents TEM picture of the eutectic phase(s) and the Mn,Fe,Si-containing particle at grain boundary of the 7075-ScZr alloy. Eutectic boundary phases and their composition in the alloy containing Sc–Zr addition were determined by EDS match as the $MgZn_2$ phase and S-phase (Al_2CuMg). Although electron diffraction (ED) was not sufficiently conclusive, it can be assumed in the light of our previous research in the hot-deformed 7075(-ScZr) alloys (see References [15,20]) that the Mn,Fe,Si-containing particle is very likely the primary cubic λ-Al(Mn,Fe,Si) phase. The presence (indirectly also the composition) of the eutectic phases was confirmed by XRD, see Figure 3. In these measurements another phase was proved: T-phase (the phase belongs to the $Al_2Zn_3Mg_3$ or $Mg_{32}(Al,Cu,Zn)_{49}$ structural type of phases). The same eutectic phase in the as-cast and hot-deformed state of the Al–Zn–Mg–Cu-based alloys was observed—see References [15,19,20]. With respect to the TEM and EDS results, it can be assumed that the $MgZn_2$ phase is also distinguishable in the XRD results, although the peaks of this phase are very weak and may be affected by the background. In view of results obtained by XRD, it can be said that the content of primary Mn,Fe,Si-containing particles observed by TEM may be relatively small and inhomogeneous.

The Sc,Zr-containing particles that correspond to the primary phase were also revealed in the 7075-Sc,Zr alloy in the HT state in several grains (Figures 1 and 4). Although the structure of the Sc(Zr)-containing particles is greatly dependent on the composition (and heat treatment routes), very similar structure (square and polygonal shapes) of the primary Sc,Zr-containing particles was detected in the Al–Mg–Sc-based [45–47] and Al–Zn–Mg–Sc-based alloys [15,19,20,48]. Figure 4 shows an AFM image according to the EDS observation of a selected primary Al_3(Sc,Zr)-phase particle in the 7075-ScZr material. In contrary to the primary Al_3Sc particles (References [45–47]), where a deep hole in the center surrounded by an area of several shell-like layers characterized by variant depths was proved, the incoherent Sc,Zr-containing particles detected here have heights falling in the interval of about 200–300 nm. Thus it can be concluded that significant altitude difference of the layers of the primary Sc-containing particles (References [45–47]), the inner central pothole, the less deep profound interlayer, the same profound of the inner convex layer around the pothole, and the marginal layer can be induced by different resistance of various layers of the primary particles to the electro-polishing before AFM observation. The surface of specimens for AFM observations was prepared only by a standard metallographic mechanical polishing and no electro-polishing was applied. It is therefore questionable whether electro-polishing connected with possible selected erosion in the sample preparation for the characterization of the Al-based materials with Sc–Zr addition by AFM should be avoided. However, this conclusion would require a more detailed examination which is not the subject of this work.

Formation of the coherent secondary Al_3(Sc,Zr) particles ($L1_2$/cP4 structure) with the size of 10–50 nm predominantly situated in grain interiors and denuded zones created along boundaries of the grains were also revealed in the heat-treated 7075-ScZr alloy (Figure 5). The particle dispersion shows a typical contrast of "the coffee bean". ED pattern of the $L1_2$-structured precipitates is shown as inset in Figure 5. No precipitates other than the Al_3(Sc,Zr) were observed inside grains by TEM, which means that other phases (if any) inside grains have very low volume fraction. Compared to the as-cast state of the alloys (see Reference [19]), it means that the high-temperature annealing (isothermal annealing at 470 °C/60 min) is sufficient to dissolve the precipitates from the Al–Zn–Mg–Cu system inside grains, but at the same time the particles from the Al–Sc–Zr system can precipitate.

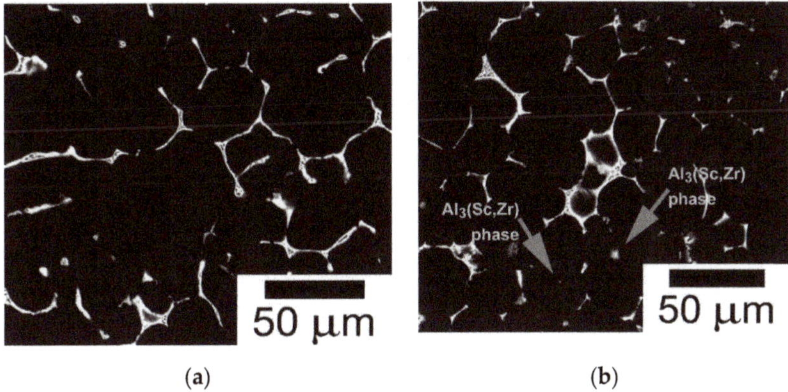

Figure 1. Microstructure overview (SEM images) of the HT state: (**a**) The 7075 alloy; (**b**) the 7075-ScZr alloy.

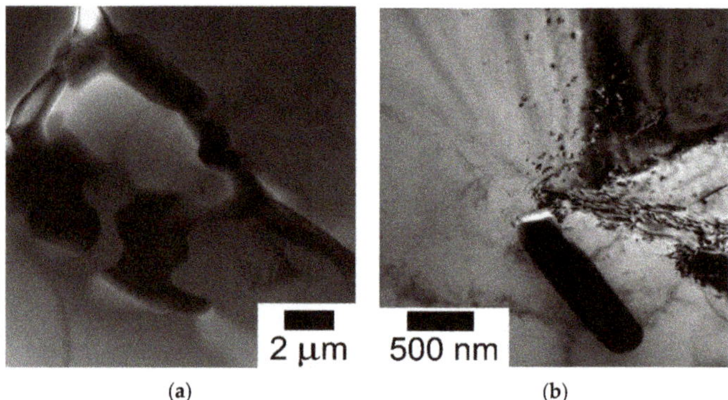

Figure 2. TEM image of the heat-treated 7075-ScZr alloy: (**a**) Eutectic phase(s); (**b**) coarse Mn,Fe,Si-containing particle.

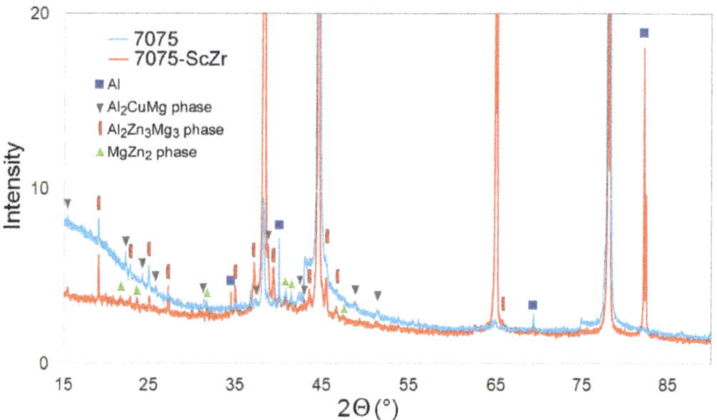

Figure 3. XRD diffraction patterns of the heat-treated 7075 and 7075-ScZr alloys with identified phases.

Figure 4. AFM images of the incoherent primary Sc,Zr-containing particle: (**a,b**) 3D topography; (**c**) phase shift image and (**d**) the height profile in the middle section of the particle (cross-section).

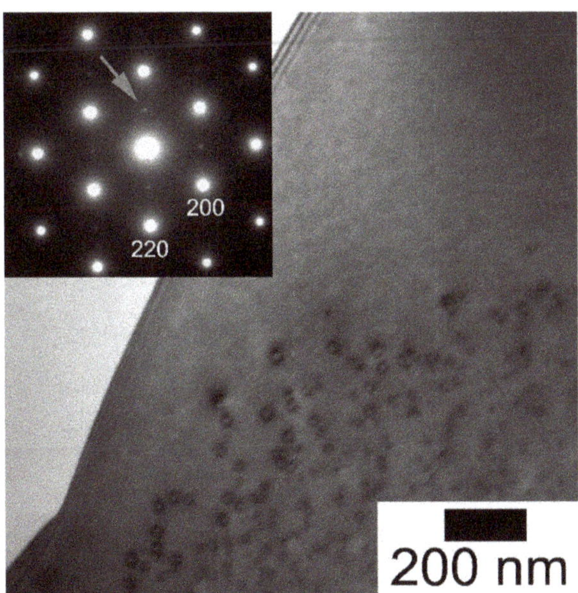

Figure 5. Microstructure image (TEM, [110]$_{Al}$ orientation) of the coherent secondary Al$_3$(Sc,Zr) particles and denuded zones in the 7075-ScZr alloy, ED pattern in the inset.

3.2. Natural Ageing of the Heat-Treated Alloys

Time evolution of the mean LT of positrons and microhardness ΔHV0.5 values of the alloys during natural ageing is shown in Figure 6. The mean LT behavior (Figure 6a) is comparable for both alloys. It can be divided into three individual stages: Up to 500 min an LT increase (stage I), then from 500 min up to ~1980 min (33 h) a roughly constant values of the LT (stage II), and above ~33 h a decrease of the LT (stage III). It can be also seen that the LT values for the 7075-ScZr alloy are higher than those for the 7075 alloy without Sc–Zr addition. Microhardness HV0.5 values (Figure 6b) increase immediately from the beginning of the natural ageing. The initial absolute HV0.5 value of the 7075 alloy (HV0.5 ≈ 70) is significantly lower than the 7075-ScZr alloy (HV0.5 ≈ 95). The microhardness values of the alloys reflect the Sc–Zr addition. This can be also seen from Table 3, where the microhardness HV0.5 values and absolute electrical resistivity ρ values (measured at temperature of liquid N$_2$—78 K) after different heat treatment are given: The state after heat treatment at 470 °C/240 min and the HT state (470 °C/60 min); and both states after the subsequent natural ageing (33,450 and 3500 h). Presence of the secondary phase from the Al–Sc–Zr-based alloys caused typical hardening effect as ΔHV ≈ 20–40, see References [15,19,20,25]. Comparable strengthening caused by co-presence of the primary incoherent and secondary coherent Sc,Zr-containing particles was observed in the deformed Al–Zn–Mg–Cu-based alloys with Sc–Zr addition [15,19,20]. Thus the presence of these Sc,Zr-containing particles is a probable reason for higher microhardness HV0.5 of the studied 7075-ScZr alloy. Table 3 also confirmations that no significant differences are between the values for the comparable states (cf. the state after annealing at 470 °C/240 min and the HT state). A difference in microhardness ΔHV0.5 values and resistivity Δρ values between the 7075 and 7075-ScZr alloys is similar within the error through the natural ageing regardless of the annealing time at 470 °C. According to the microstructure observation of the alloys (see above), this also justifies the claim that the heat treatment at 470 °C/60 min is sufficient for the Al–Zn–Mg–Cu system.

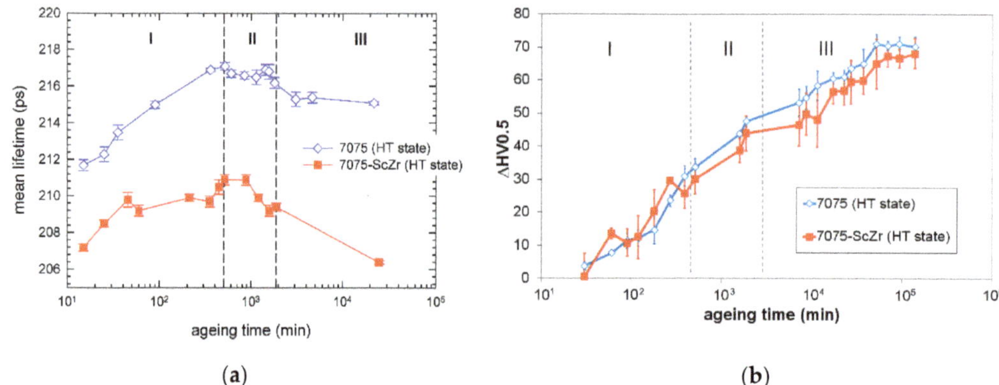

(a)   (b)

Figure 6. The evolution during natural ageing of the 7075 and 7075-ScZr alloys: (**a**) Mean LT values; (**b**) microhardness changes ΔHV. Vertical dashed lines indicate individual stages I–III.

Table 3. Values of microhardness HV0.5 and electrical resistivity ρ in the different state of the 7075 and 7075-ScZr alloys after different heat treatment and during subsequent natural ageing (NA) after 33, 450, and 3500 h.

Alloy	Heat Treatment	HV0.5				ρ (n$\Omega\cdot$m)		
		Initial State	NA 33 h	NA 450 h	NA 3500 h	Initial State	NA 33 h	NA 3500 h
7075	HT state	70 ± 3	117 ± 2	134 ± 3	145 ± 2	27 ± 1	35 ± 1	36 ± 1
7075	470 °C/240 min	66 ± 6	115 ± 2	134 ± 2	144 ± 4	-	-	-
7075-ScZr	HT state	95 ± 4	139 ± 5	155 ± 6	168 ± 3	26 ± 1	33 ± 1	34 ± 1
7075-ScZr	470 °C/240 min	94 ± 4	138 ± 5	154 ± 4	167 ± 2	-	-	-

By two exponential components the LT spectra of all samples studied can be well described. Lifetimes τ_1 and τ_2 of the constituents and relative intensity I_2 of the longer constituent are plotted in Figure 7a,b, respectively. The shorter constituent with the lifetime τ_1 comes from free positrons (not captured in defects). The longer constituent with $\tau_2 \approx 220$ ps can be ascribed to the positrons captured at solute (co-)clusters in the alloys [20]. The lifetime τ_2 is comparable in both 7075 and 7075-ScZr alloys. This indicates that both alloys contain similar type of solute (co-)clusters. Note that LT component with similar lifetime τ_2 was observed also in these alloys subjected to hot deformation [20]. Our previous CDB investigations revealed that these clusters contain mainly Mg and Zn solutes [20]. It also has to be mentioned that coherent secondary as well as primary Al$_3$(Sc,Zr) particles do not trap positrons [19,20] but the Sc–Zr addition in the alloys can affect the development of the other phases (e.g., non-eutectic phases of the Al–Zn–Mg-based system [20]).

Positron annihilation spectroscopy (PAS) investigations of the Al–Zn–Mg-based alloys indicated that the interaction between solute atoms and vacancies has an influence on (co-)clusters/GP zones [19,20,25]. Parallel to PAS observations in the Al–Mg–Si-based [49–51] and Al–Zn–Mg–Cu-based alloys [19,20] one can conclude that immediately after high temperature treatment (solution treatment) the studied alloys contain vacancies associated with single and/or multiple Zn- or Mg-solutes and/or Zn,Mg(-co)-clusters developed during quenching. These solute (co-)clusters further evolve in the course of natural ageing. Owing to the strengthening effect of the (co-)clusters, the microhardness HV0.5 of the 7075 and 7075-ScZr alloys increases continuously during ageing at RT which is probably due to growing and development of these solute (co-)clusters in analogy with Al–Zn–Mg-based alloys (e.g., References [25,52,53]).

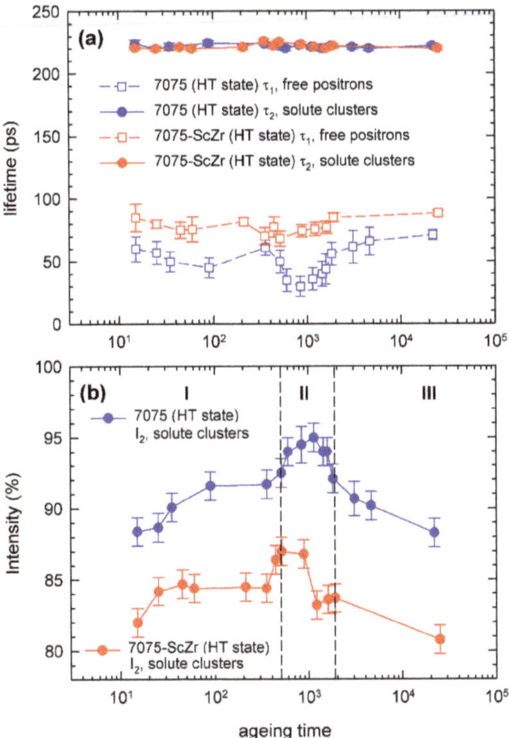

Figure 7. The evolution of: (**a**) The lifetimes τ_1 and τ_2; (**b**) the intensity I_2 of the HT state of the alloys subjected to natural ageing.

The intensity I_2 in the 7075-ScZr sample is lower than in the 7075 alloy. It indicates that the Sc–Zr addition has a slight negative influence on the concentration of solute (co-)clusters. Si, Cu or Mg solutes are most probably bound to Sc and Zr solutes and/or to the (secondary) $Al_3Sc(Zr)$ particles. This presumed conclusion is further supported by the fact that that the 7075-ScZr sample, i.e., the alloy with lower concentration of solute (co-)clusters, also exhibits slightly lower changes in microhardness HV0.5, see Figure 6b. However, it is also not possible to eliminate mutually different supersaturation in the solid solution.

The majority of positrons is trapped in solute (co-)clusters ($I_2 > 80\%$) already after the heat treatment at ~470 °C. It means that solute (co-)clusters were formed already during quenching of the sample. In the course of natural ageing the concentration of solute clusters increases in the stage I. This is reflected by an increase of I_2 in the stage I which reaches maximum in the stage II. A long term natural ageing (stage III) leads to a decrease of I_2 due to agglomeration of solute (co-)clusters into bigger objects ((precursors of the) GP zones). This results in an increase of the mean distance between clusters and thereby a decrease of their density.

Taking into account the behavior of the constituent (especially Zn, Mg, and Cu) at temperatures from RT to ~80 °C [20] and the calculated contribution of the solute concentration to the resistivity of Al (e.g., References [1,54,55]), the resistivity development during natural ageing is mainly associated with the change of Zn and Mg concentrations in the Al matrix. In the first approximation it can be assumed that the (co-)clusters/GP zones are composed of Zn and Mg atoms in a ratio of: (a) 1:1; (b) 1:2, see e.g., References [10,15,17,19–33,56–58]. If we assume this, then the resistivity $\Delta\rho$ changes in the HT + NA state (after natural ageing up to 3500 h), $\Delta\rho \approx 8$ n$\Omega\cdot$m, see Table 3, would correspond to the decrement of Zn- and Mg-concentration in the matrix: (a) $\Delta C_{Zn,Mg} \approx 0.7 \pm 0.1$ at.%; (b) $\Delta C_{Zn} \approx 0.8 \pm 0.2$ at.% and $\Delta C_{Mg} \approx 0.4 \pm 0.1$ at.%. This decrement reasonably implicates that the changes in resistivity, microhardness and PAS characteristics can be attributed to the process of the formation/coarsening of the (co-)clusters/GP zones. These conclusions suitably complement the findings obtained through PAS.

In addition to microstructural characterization during the natural ageing, the corrosion characteristics of the alloys in the HT + NA state (natural ageing after 3500 h) were studied. The results of the potentiodynamic polarization are shown in Table 4 and Figure 8. The 7075-ScZr alloy has a higher corrosion potential E_{corr} than the 7075 alloy without Sc–Zr addition. A comparable difference of the $\Delta E_{corr} \approx 28$ mV caused by the Sc–Zr addition was measured in our previous study in the hot-rolled alloys with the comparable composition, see Reference [15]. In the studied 7075 and 7075-ScZr alloys the passivation region, where the corrosion current kept almost the same value ΔE_{corr} (see Table 4), was observed, which means that passive films were formed on the surfaces. A lower value of corrosion current density I_{corr} in the 7075-ScZr alloy (2.23 μA/cm^2) indicates a better corrosion resistance and a slower corrosion rate (~0.0243 mm/a) than for the 7075 alloy (2.96 μA/cm^2 and ~0.0322 mm/a). For the studied alloys it can be concluded that the positive influence on corrosion properties is (probably) mainly caused by the Sc–Zr addition. Although corrosion measurement results can be found on many Al-based alloys, there are really very few detailed electrochemical measurements on the Al(–Sc)–Zr-based alloys (e.g., References [41,42,59–62]) and even less on the alloys of this type (7xxx series) with Sc and Zr elements in the available literature. In addition to our previous study in the hot-rolled alloys with comparable composition [15], the observed results are in agreement with Reference [62] where better corrosion resistance (especially slower corrosion rate and lower corrosion current density) was measured by electrochemical measurements for the AlZnMgCuScZr alloy than for the AlZnMgCu alloy in the EXCO testing solution.

Table 4. Corrosion characteristics of the HT + NA (3500 h) alloys obtained by potentiodynamic polarization measurements (measured at RT) in the 0.1 M NaCl solution. WL represents weight loss, CR corrosion rate, I_{corr} corrosion current density and E_{corr} corrosion potential.

Alloy	I_{corr} (μA/cm^2)	WL (g/m^2/d)	CR (mm/a)	E_{corr} (mV)
7075	2.96	~0.238	~0.0322	−1154
7075-ScZr	2.23	~0.180	~0.0243	−1125

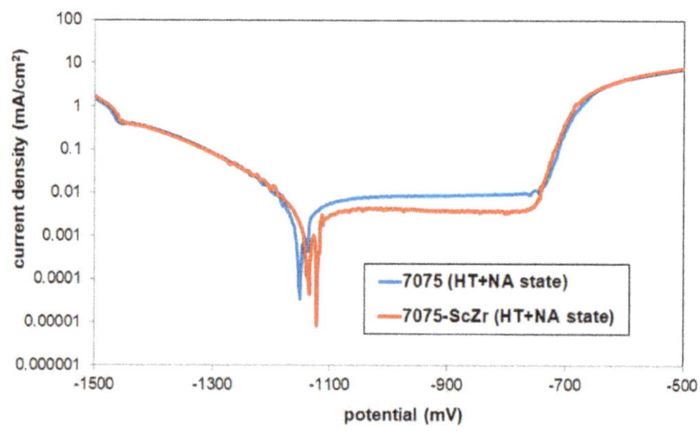

Figure 8. Potentiodynamic polarization diagram (measured at RT) in the 0.1 M NaCl solution.

3.3. Phase Development during Isochronal Annealing

The ageing processes in the Al–Zn–Mg–Cu-based alloys are complex and the decomposition of supersaturated solid solutions obtained by quenching takes place in several formation steps [15,19,20,24,63,64]. Typically, the precipitation of the (co-)clusters, precursors of the GP zones, and coherent GP zones precedes the formation of the semicoherent intermediate precipitates and incoherent equilibrium precipitates during isochronal annealing [15,19,20,25,65–67]. The early precipitation stages in these alloys are abundant and can have a significant influence the resistivity and (micro)hardness development during the natural ageing and/or the beginning of the annealing [25,30,49,52,53]. Isochronal resistivity ($\Delta\rho/\rho_0$) and microhardness annealing curves of the HT alloys are presented in Figure 9. Electrical resistivity decrease in two stages (1-stage up to ~140 °C and 2-stage between 140 °C and 280 °C). Then the electrical resistivity is nearly constant (3-stage) and after annealing at temperatures higher than ~360 °C (4-stage) increases insignificantly. 1- and 2-stages of the resistivity decrease are connected with a maximum hardening HV0.5 after annealing up to 180–220 °C (Figure 9). The higher HV0.5 values of the 7075-ScZr in the HT state is probably caused by the secondary/primary Al$_3$(Sc,Zr)-phase particles (already existed in the HT state—see Figures 4 and 5). The comparable effect on HV0.5 values due to the presence of these particles is observed during natural ageing of the alloys (see Table 3).

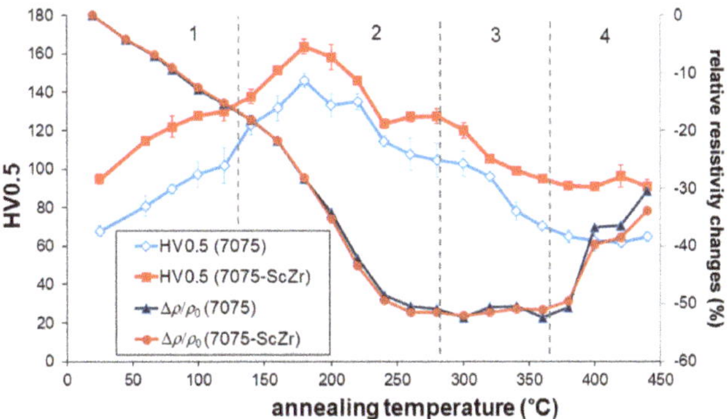

Figure 9. Isochronal annealing curves of relative resistivity changes and microhardness HV0.5 changes with standard deviation of the HT alloys.

The microhardness isochronal curves for the HT + NA (natural ageing for 3500 h) alloys and the 7075-ScZr alloy heat treated at 470 °C/240 min after natural ageing (NA) for 3500 h are given in Figure 10. The microhardness values of the 7075-ScZr alloys are comparable for both alloys (heat treatment at 470 °C/60 min and 470 °C/240 min) after natural ageing. This indicates that the annealing at 470 °C/60 min is sufficient for the Al–Zn–Mg–Cu system, as stated above. The initial microhardness HV0.5 values of the HT + NA alloys in comparison with the HT alloys without NA (compare Figures 9 and 10) reflects the presence of the (co-)clusters/precursors of the GP zones developed during natural ageing. The HV0.5 values decrease to a local minimum at ~90 °C. The temperature range of the hardening peak can be observed at 150–210 °C. It is seen that the Sc–Zr addition has almost no effect on microhardness changes at 210–330 °C. But after that the HV values of the alloy with the Sc–Zr addition after only a slight decrease reach constant course in contrast to the continual decrease of the 7075 alloy. The ΔHV0.5 difference between 7075 and 7075-ScZr alloys at the end of annealing is nearly HV0.5 ≈ 20 microhardness development of the alloys studied reflects the Sc–Zr addition, again.

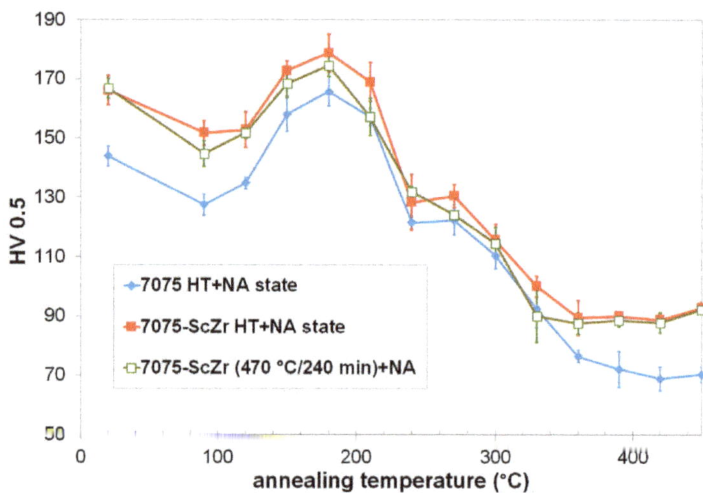

Figure 10. Isochronal annealing curves of microhardness HV0.5 changes with standard deviation of the HT + NA alloys and 7075-ScZr alloy heat treated at 470 °C/240 min after natural ageing for 3500 h.

Out of the comparison of the microhardness and electrical resistivity changes annealing curves of the studied materials up to ~140 °C (Figures 9 and 10) it could be assumed that the precursors of the GP zones and/or GP zones are formed in the HT alloys during the isochronal annealing. This process is connected with microhardness and electrical resistivity increase at the 1-stage. Despite this fact, the precursors of the GP zones and/or GP zones are dissolved first in the HT + NA alloys. This dissolution leads to the microhardness decrease (up to ~100 °C). The results are in an agreement with the measurements of the Al–Zn–Mg(–Sc–Zr) alloys in our previous study (see Reference [25]).

Figure 11 shows microstructure of the 7075-ScZr alloy in the HT state (without NA) after isochronal heat treatment up to 220 °C, where the dense particle dispersion can be seen. Thus the main resistivity decrease (2-stage) and microhardness increase (140–220 °C) in the alloys studied are due to the precipitation of these particles. According to the literature (e.g., References [15,19,20,31–33]), the following phases in this temperature range come into consideration: (a) Metastable η'-phase (hexagonal, $a = 0.496$ nm, $c = 0.935$ nm or bcc, $a = 1.422$ nm, MgZn$_2$) [19,20,25,32,33]), (b) non-eutectic stable T-phase (bcc, $a = 1.435$ nm, Al$_2$Zn$_3$Mg$_3$ and/or Mg$_{32}$(Al$_{1-x}$Zn$_x$)$_{49}$) [20,25,31], (c) GP zones (hexagonal, $a = 0.496$ nm, $c = 0.935$ nm or bcc, $a = 1.422$ nm) [19,20,25,31,32]. Figure 12 shows the diffraction patterns from different grains of the 7075-ScZr alloy after isochronal heat treatment up to 220 °C taken near: (a) [110]$_{Al}$ and (b) [111]$_{Al}$ orientations. Figure 12a contains weak spots assigned to the secondary Al$_3$(Sc,Zr)-phase particles, GP zones and non-eutectic T-phase particles. This is in conformity with recently published results, see References [19,20,31,32,58]. On the contrary, Figure 12b also shows weak spots which can be assigned to metastable η'-phase particles (see References [25,32,33]) and to the GP zones and secondary Al$_3$(Sc,Zr)-phase particles [19,25,57,67]. Although the presence of the GP zones, T-, η'- and Al$_3$(Sc,Zr)-phase particles has been demonstrated in electron diffraction (Figure 12), the individual types of the particles could not be easily recognized in conventional TEM image thanks to the high density of the particles (Figure 11). One can therefore mention that the phase composition is different in different places of the samples even after heat treatment.

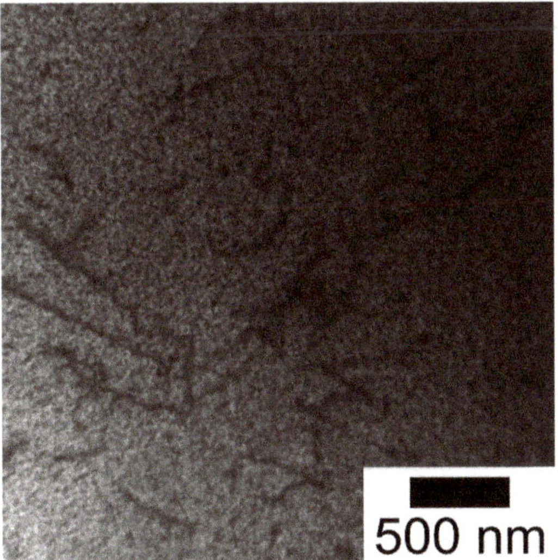

Figure 11. Microstructure (TEM) of the 7075-ScZr alloy (near [110]$_{Al}$ orientation) annealed up to 220 °C.

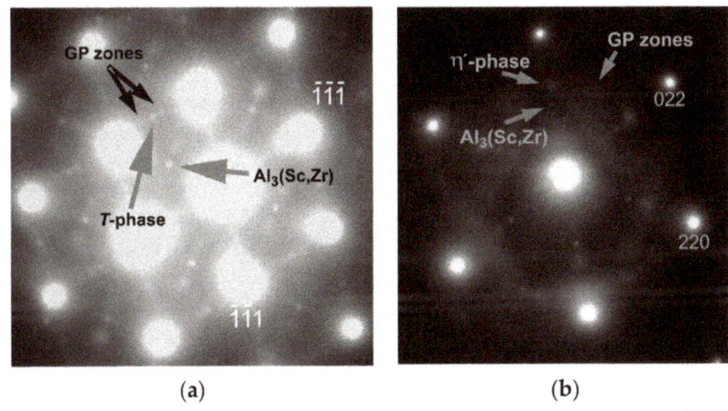

(a)   (b)

Figure 12. SAED pattern of the 7075-ScZr alloy after isochronal heat treatment up to 220 °C: (**a**) Near [110]$_{Al}$ orientation with weak spots from the GP zones, *T*-phase and Al$_3$(Sc,Zr) phase; (**b**) near [111]$_{Al}$ orientation with weak spots from the GP zones, η'-phase and Al$_3$(Sc,Zr) phase.

SEM and TEM observation of the 7075-ScZr alloy in the HT state after isochronal heat treatment up to 440 °C proved a coarsened secondary Sc,Zr-containing phase particles—see Figure 13. A volume fraction and composition of the eutectic grain boundary phase stay unmodified and the grain size does not vary in comparison to the initial HT state of the alloys (see above Chapter 3.1.). Furthermore, the Zn,Mg-containing (co-)clusters/GP zones and other phases (*T*-phase, η'/η-phase) disappeared during the isochronal annealing up to this temperature. The dissolution of these phases is mainly connected with the resistivity increase above ~360 °C (see Figure 9). At temperatures above ~240 °C, particle precipitation of the η-phase and *S*-phase was proved in the Al–Zn–Mg–Cu-based systems, e.g., see References [20,25]. This precipitation is a probable explanation of the undulating of the electrical resistivity curves in the 3-stage (Figure 9) and microhardness HV0.5 curves (Figures 9 and 10) at a temperature range of 250–350 °C. One can also conclude that the annealing is sufficient for homogenization of the Al–Zn–Mg–Cu-based system, but it is insufficient for the Al–Sc–Zr-based system, again. Microhardness development of the all alloys studied (Figures 9 and 10) reflects the Sc–Zr addition at temperatures above ~300 °C. Assuming the presence of the primary and secondary Al$_3$(Sc,Zr)-phase particles it can be also concluded that the yield strength cannot be captured as a linear superposition of several strength contributions. This conclusion is in an agreement with other observations in commercial 2xxx, 6xxx and 7xxx alloys, see References [20,25,65,67–69].

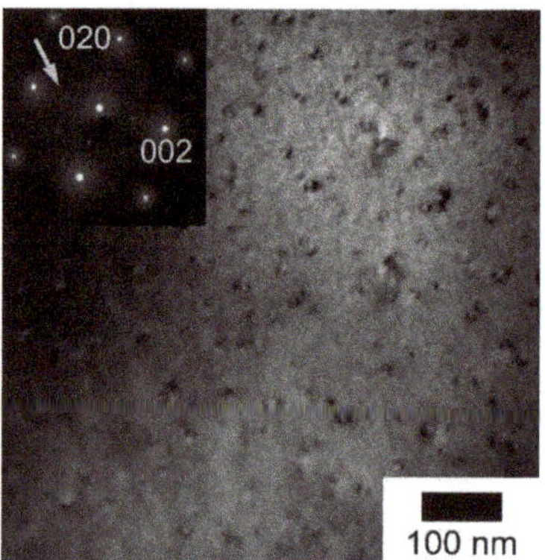

Figure 13. Microstructure (TEM) of the 7075-ScZr alloy after isochronal annealing up to 440 °C: Particles of the secondary Sc,Zr-containing phase with L1$_2$ structure (ED pattern of [100]$_{Al}$ zone in the inset).

3.4. Thermal Properties of the Alloys

Due to the acquired additional characteristics about the kinetics of phase development in the 7075(-ScZr) alloys in the different states, the apparent activation energy of the thermal processes by using DSC were calculated. The following states of the 7075 and 7075-ScZr alloys were used for the thermal study: HT, HT + NA (3500 h at RT) and HT + NA15000 (15,000 h at RT). Figure 14 shows, for clarity only, selected DSC thermographs up to 400 °C for one heating rate (5 °C/min). The curves mainly differ due to natural ageing (the HT state vs. HT + NA state and HT + NA15000 state). One endothermic effect marked B and five exothermic effects marked A, C–F (see Figure 14) are shown in the DSC curves of the samples.

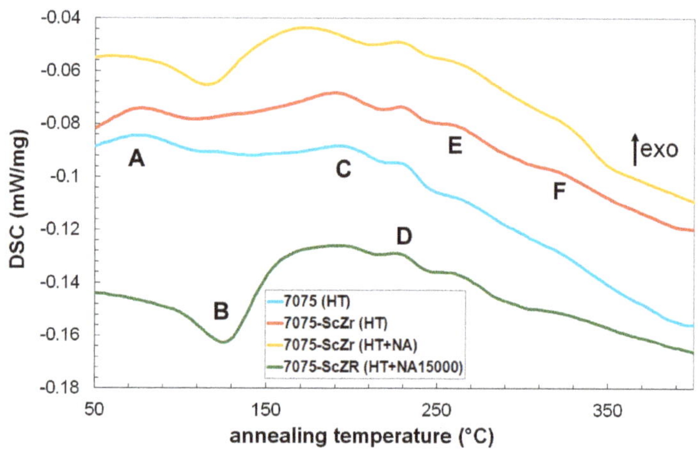

Figure 14. Selected DSC curves (5 °C/min) of the 7075 alloy in the HT state and the 7075-ScZr alloy in the HT, HT + NA, and HT + NA15000 state.

Table 5 shows characteristic temperatures T_m of a DSC peak of maxima/minima of the thermal effects A–E in the alloys studied in the HT and HT + NA state. The exothermic effect C in the HT + NA state and the effect F in all the alloys studied were very often too weak to be analyzed through the materials studied, so they are not listed in Table 5.

Table 5. Characteristic temperatures T_m of a DSC peak of maxima/minima of the thermal effects A–E of the 7075 and 7075-ScZr alloys in the HT and HT + NA state.

Effect/Alloy	1 °C/min	2 °C/min	5 °C/min	10 °C/min	20 °C/min
Effect A (7075 HT)	58	63	72	83	90
Effect A (7075-ScZr HT)	58	63	72	84	91
Effect B (7075 HT + NA)	106	110	118	126	141
Effect B (7075-ScZr HT + NA)	104	110	118	127	142
Effect C (7075 HT)	165	177	191	201	209
Effect C (7075-ScZr HT)	165	178	189	198	-
Effect D (7075 HT)	208	219	231	239	255
Effect D (7075-ScZr HT)	207	219	230	241	255
Effect D (7075 HT + NA)	207	217	231	242	249
Effect D (7075-ScZr HT + NA)	208	216	229	239	250
Effect E (7075 HT)	237	255	263	277	295
Effect E (7075-ScZr HT)	237	254	265	276	294
Effect E (7075 HT + NA)	240	254	267	277	298
Effect E (7075-ScZr HT + NA)	238	256	265	277	292

The apparent activation energies Q of the observed effects A–E (see Table 6) were determined by the Kissinger method [24,70] based on the data obtained from the DSC curves (Table 5) and the knowledge that the characteristic temperatures T_m correspond to the maxima (minima) of the observed process and maximum process speed of the observed effect, respectively. Then the activation energies Q for the individual thermal effects are determined by a linear regression, i.e., $\ln(\delta/T_m^2)$ plotted vs. $1/T_m$, where δ is the heating rate. Some activation energies were not calculated as a consequence of a low markedness of the maximum heat flow (e.g., effect C for the HT + NA alloys).

Table 6. Activation energies Q of the thermal effects A–E determined by the Kissinger analysis of DSC data.

Sample	Effect (Process)	Activation Energy Q (kJ/mol)
7075 HT	A (GP zones formation)	84 ± 6
	C (η'-phase precipitation)	111 ± 6
	D (T-phase precipitation)	130 ± 11
	E (η-phase precipitation)	122 ± 9
7075-ScZr HT	A (GP zones formation)	81 ± 5
	C (η'-phase precipitation)	113 ± 10
	D (T-phase precipitation)	127 ± 10
	E (η-phase precipitation)	124 ± 10
7075 HT + NA	B (GP zones dissolution)	103 ± 13
	D (T-phase precipitation)	136 ± 5
	E (η-phase precipitation)	122 ± 12
7075-ScZr HT + NA	B (GP zones dissolution)	100 ± 11
	D (T-phase precipitation)	137 ± 5
	E (η-phase precipitation)	130 ± 12

The exothermic effect A corresponds well to the hardening and resistivity decrease (1-stage) up to ~140 °C of the HT alloys, see Figure 9. The endothermic effect B corresponds well to the softening up to ~120 °C of the HT + NA alloys (Figure 10). The effect A obviously corresponds to the production of (the precursors of) the GP zones, effect B to their dissolution, respectively. Given the available experimental options, it is not possible to distinguish between coarsening of the precursors of the GP zones ((co-)clusters) and the GP zones. The activation apparent energy values of the effect B (~102 kJ/mol) are in a good agreement for the dissolution of the GP zones, those reported in References [25,71–73]. The temperature range of the dissolution is logically higher and activation energy values are logically also higher than the values for the effect A (~82 kJ/mol)—the formation of the (co-)clusters/GP zones [25,52,63]. The temperature range of the effect B (i.e., (co-)clusters/GP zones dissolution) is slightly influenced by the natural ageing time, cf. HT + NA state and HT + NA15000 state in Figure 14. However, measurements on the HT + NA15000 sample were made only once (at one heating rate only).

The exothermic processes C and D correspond well to the hardening with a maximum at ~180 °C of the alloys in the HT and HT + NA state as well as the resistivity decrease in the 2-stage (cf. Figure 9, Figure 10, and Figure 14). As the microscopy observation by TEM at 220 °C is shown in Figures 11 and 12, these effects C and D can be assumed to be associated with metastable η'-phase and non-eutectic stable T-phase particle precipitation. The activation energy of the effects C and D were calculated as ~112 kJ/mol and ~133 kJ/mol, respectively. The calculated value for the precipitation of the η'-phase particles in the studied alloys excellently agrees with the values for the precipitation of the η'-phase determined in the Al–Zn–Mg-based alloys [25,72]. Although the data in literature for the T-phase precipitation are usually not available, in our previous study [25] we calculated the activation energy in the alloys with comparable composition without Cu for this precipitation in the range of 128–168 kJ/mol. But the result was based only on a few experimental points. Given the direct evidence using TEM (Figure 12) in this study, it can be concluded that the effect D is very probably connected with the T-phase precipitation. From the obtained results it can also be deduced that the formation of the η'-phase particles is affected by natural ageing at RT similarly to the formation/dissolution of the (co-)clusters/GP zones, while the T-phase precipitation is not affected by natural ageing.

The activation energy for the effect E is calculated as ~125 kJ/mol. The energy value could not be determined for the effect F because a weak thermal process was detected only for some heating rates. The calculated value of the effect E is in an agreement with literature for the precipitation of the stable η-phase: 118–142 kJ/mol, see References [25,71–73]. Although there is no direct microscopic evidence, the temperature region of the effect F corresponds very well to the temperature range (240–300 °C) of the precipitation of irregular shape S-phase particles [20,25,58,74,75]. Conversely, formation of the η-phase particles can occur below ~240 °C in the Al-based alloys [19,20]. Although the data in the literature for this precipitation are insufficient, the calculated values indicate that precipitation of the T-phase takes place by a different mechanism than the precipitation of the stable η-phase. One can also conclude that Sc–Zr addition has a little effect on the ongoing transformation processes, probably due to the fact that these elements are not in solid solution, but in the Al$_3$(Sc,Zr)-phase particles.

4. Conclusions

A detailed characterization of the phase development and age hardening of the heat-treated commercial aluminum 7075(Al–Zn–Mg–Cu–Mn–Si–Fe(–Sc–Zr))-based alloys can be stated in the following points:

- The microstructure of the initial (heat-treated) state of the alloys is very complex. The observation proved that the eutectic phases consist of four types at ones: MgZn$_2$ phase, Al$_2$CuMg phase (S-phase), Al$_2$Zn$_3$Mg$_3$/Mg$_{32}$(Al,Cu,Zn)$_{49}$ phase (T-phase) and primary cubic λ-Al(Mn,Fe,Si) phase. In addition, two types of non-eutectic particles in the alloys with Sc–Zr addition are present: Primary incoherent Al$_3$(Sc,Zr) particles with square and polygonal shapes and secondary coherent Al$_3$(Sc,Zr) particles.
- The Sc,Zr-containing alloys cannot be completely homogenized due to these elements. However, for the 7xxx-based system, a heat treatment above ~440 °C is sufficient (unless eutectic phase is considered). It is not even possible to observe the differ-

ence in characterization of the properties after the high-temperature annealing at 470 °C/60 min and 470 °C/240 min, respectively.

- Microhardness values reflect the Sc–Zr addition in the 7075-ScZr alloy. Strengthening is caused by the presence of the primary/secondary $Al_3(Sc,Zr)$-phase particles. Positive influence on corrosion properties is also caused by the Sc–Zr addition.
- Single and/or multiple Zn- or Mg-solutes and/or Zn,Mg(-co)-clusters developed during quenching immediately after high temperature treatment in the alloys. These solute (co-)clusters further evolve in the course of natural ageing. This process causes a significant increase in microhardness and electrical resistivity values. A long term natural ageing leads to a coarsening of the solute (co-)clusters into bigger objects (probably precursors of the GP zones).
- Addition of Sc and Zr has a slight negative influence on the concentration of solute (co-)clusters. Si, Cu or Mg solutes are most probably bound to Sc and Zr solutes and/or to the $Al_3Sc(Zr)$-phase particles. However, in general it can be said that the co-presence of the Sc- and Zr- elements have only a little effect on the phase transformations of the Al–Zn–Mg–Cu system.
- Formation of the η'/η-phase is slightly suppressed in the alloys after natural ageing. The dissolution of the precursors of the GP zones/GP zones is shifted to higher temperatures depending on the time of natural ageing.
- The apparent activation energy values of the observed thermal processes were calculated as: Formation of the precursors of the GP zones/GP zones: ~82 kJ/mol, dissolution of the (co-)clusters/GP zones: ~102 kJ/mol, formation of the metastable η'-phase: ~112 kJ/mol, formation of the stable T-phase: ~133 kJ/mol, formation of the stable η-phase: ~125 kJ/mol.
- A combination of both precipitation sequences of the Al–Zn–Mg–Cu-based system was observed in the studied alloys.

Author Contributions: Conceptualization, M.V.; methodology, M.V. and V.N.; validation, J.C. and V.K.; investigation, M.V., V.K., J.C., M.L., T.K., F.L., L.B., P.K., E.C., H.K., M.C., V.S., S.Z., V.N., V.-D.N., and P.K.; writing—original draft preparation, M.V., J.C., and V.K.; writing—review and editing, M.V., V.K., J.C., and M.C.; supervision, M.V. and V.N.; project administration, M.V.; funding acquisition, M.V. All authors have read and agreed to the published version of the manuscript.

Funding: This research was funded by The Czech Science Foundation (GACR), grant number 17-17139S.

Institutional Review Board Statement: Not applicable.

Informed Consent Statement: Not applicable.

Data Availability Statement: The data used to support the findings of this study are available from the corresponding author upon request.

Acknowledgments: The authors are also grateful to Ivana Stulikova, Bohumil Smola and Petra Hoffmannova for their help.

Conflicts of Interest: The authors declare no conflict of interest. The funders had no role in the design of the study; in the collection, analyses, or interpretation of data; in the writing of the manuscript, or in the decision to publish the results.

References

1. Toropova, L.S.; Eskin, D.G.; Kharakterova, M.L.; Dobatkina, T.V. *Advanced Aluminium Alloys Containing Scandium—Structure and Properties*, 1st ed.; Gordon and Breach Science Publisher: Amsterdam, The Netherlands, 1998.
2. Roskill—Commodity Research, Consulting & Events. Available online: http://www.roskill.com/ (accessed on 2 April 2020).
3. Schuman, S.; Friedrich, F. The use of magnesium in cars—Today and in future. In *Magnesium Alloys and Their Applications*; Mordike, B.L., Kainer, K.U., Eds.; Werkstoff-Informationsgesselschaft: Frankfurt, Germany, 1998.
4. Jones, M.J.; Humphreys, F. Interaction of recrystallization and precipitation: The effect of Al3Sc on the recrystallization behaviour of deformed aluminium. *Acta Mater.* **2003**, *51*, 2149–2159. [CrossRef]
5. Dursun, T.; Soutis, C. Recent developments in advanced aircraft aluminium alloys. *Mater. Des.* **2014**, *56*, 862–871. [CrossRef]

6. Ghosh, A.; Ghosh, M. Microstructure and texture development of 7075 alloy during homogenisation. *Philos. Mag. A Mater Sci.* **2018**, *98*, 1470–1490. [CrossRef]

7. Wen, K.; Xiong, B.; Zhang, Y.; Li, Z.; Li, X.; Yan, L.; Yan, H.; Liu, H. Measurement and Theoretical Calculation Confirm the Improvement of T7651 Aging State Influenced Precipitation Characteristics on Fatigue Crack Propagation Resistance in an Al–Zn–Mg–Cu Alloy. *Met. Mater. Int.* **2019**, 1–17. [CrossRef]

8. Yuan, B.; Guo, M.; Wu, Y.; Zhang, J.; Zhuang, L.-Z.; Lavernia, E.J. Influence of treatment pathways on the precipitation behaviors of Al-Mg-Si-Cu-(Zn)-Mn alloys. *J. Alloys Compd.* **2019**, *797*, 26–38. [CrossRef]

9. Mondal, C.; Mukhopadhyay, A. On the nature of T(Al2Mg3Zn3) and S(Al2CuMg) phases present in as-cast and annealed 7055 aluminum alloy. *Mater. Sci. Eng. A* **2005**, *391*, 367–376. [CrossRef]

10. Zhang, Y.; Pelliccia, D.; Milkereit, B.; Kirby, N.; Starink, M.J.; Rometsch, P. Analysis of age hardening precipitates of Al-Zn-Mg-Cu alloys in a wide range of quenching rates using small angle X-ray scattering. *Mater. Des.* **2018**, *142*, 259–267. [CrossRef]

11. Mostafapoor, S.; Malekan, M.; Emamy, M. Thermal analysis study on the grain refinement of Al–15Zn–2.5Mg–2.5Cu alloy. *J. Therm. Anal. Calorim.* **2017**, *127*, 1941–1952. [CrossRef]

12. Shu, S.; De Luca, A.; Dunand, D.C.; Seidman, D.N. Effects of W micro-additions on precipitation kinetics and mechanical properties of an Al–Mn–Mo–Si–Zr–Sc–Er alloy. *Mater. Sci. Eng. A* **2020**, 140550. [CrossRef]

13. Vo, N.Q.; Seidman, D.N.; Dunand, D.C. Effect of Si micro-addition on creep resistance of a dilute Al-Sc-Zr-Er alloy. *Mater. Sci. Eng. A* **2018**, *734*, 27–33. [CrossRef]

14. Li, H.; Gao, Z.; Yin, H.; Jiang, H.; Su, X.; Bin, J. Effects of Er and Zr additions on precipitation and recrystallization of pure aluminum. *Scr. Mater.* **2013**, *68*, 59–62. [CrossRef]

15. Kodetová, V.; Vlach, M.; Kudrnová, H.; Leibner, M.; Málek, J.; Cieslar, M.; Bajtošová, L.; Harcuba, P.; Neubert, V. Annealing effects in commercial aluminium hot-rolled 7075(–Sc–Zr) alloys. *J. Therm. Anal. Calorim.* **2020**, *142*, 1613–1623. [CrossRef]

16. Glerum, J.A.; Kenel, C.; Sun, T.; Dunand, D.C. Synthesis of precipitation-strengthened Al-Sc, Al-Zr and Al-Sc-Zr alloys via selective laser melting of elemental powder blends. *Addit. Manuf.* **2020**, *36*, 101461. [CrossRef]

17. Zhang, M.; Liu, T.; He, C.; Ding, J.; Liu, E.; Shi, C.; Li, J.; Zhao, N. Evolution of microstructure and properties of Al–Zn–Mg–Cu–Sc–Zr alloy during aging treatment. *J. Alloys Compd.* **2016**, *658*, 946–951. [CrossRef]

18. Emani, S.V.; Benedyk, J.; Nash, P.; Chen, D. Double aging and thermomechanical heat treatment of AA7075 aluminum alloy extrusions. *J. Mater. Sci.* **2009**, *44*, 6384–6391. [CrossRef]

19. Vlach, M.; Čížek, J.; Kodetová, V.; Kekule, T.; Lukáč, F.; Cieslar, M.; Kudrnová, H.; Bajtošová, L.; Leibner, M.; Harcuba, P.; et al. Annealing Effects in Cast Commercial Aluminium Al–Mg–Zn–Cu(–Sc–Zr) Alloys. *Met. Mater. Int.* **2019**, 1–10. [CrossRef]

20. Vlach, M.; Čížek, J.; Kodetova, V.; Leibner, M.; Cieslar, M.; Harcuba, P.; Bajtosova, L.; Kudrnova, H.; Vlasak, T.; Neubert, V.; et al. Phase transformations in novel hot-deformed Al–Zn–Mg–Cu–Si–Mn–Fe(–Sc–Zr) alloys. *Mater. Des.* **2020**, *193*, 108821. [CrossRef]

21. Rometsch, P.A.; Zhang, Y.; Knight, S. Heat treatment of 7xxx series aluminium alloys—Some recent developments. *Trans. Nonferr. Met. Soc. China* **2014**, *24*, 2003–2017. [CrossRef]

22. Rout, P.K.; Ghosh, M.; Ghosh, K. Microstructural, mechanical and electrochemical behaviour of a 7017 Al–Zn–Mg alloy of different tempers. *Mater. Charact.* **2015**, *104*, 49–60. [CrossRef]

23. Feng, C.; Shou, W.-B.; Liu, H.; Yi, D.; Feng, Y.-R. Microstructure and mechanical properties of high strength Al–Zn–Mg–Cu alloys used for oil drill pipes. *Trans. Nonferr. Met. Soc. China* **2015**, *25*, 3515–3522. [CrossRef]

24. Starink, M.; Wang, S. A model for the yield strength of overaged Al–Zn–Mg–Cu alloys. *Acta Mater.* **2003**, *51*, 5131–5150. [CrossRef]

25. Vlach, M.; Kodetová, V.; Smola, B.; Čížek, J.; Kekule, T.; Cieslar, M.; Kudrnová, H.; Bajtošová, L.; Leibner, M.; Procházka, I. Characterization of phase development in commercial Al-Zn-Mg(-Mn,Fe) alloy with and without Sc,Zr-addition. *Met. Mater.* **2018**, *56*, 367–377. [CrossRef]

26. Priya, P.; Johnson, D.R.; Krane, M.J. Precipitation during cooling of 7XXX aluminum alloys. *Comput. Mater. Sci.* **2017**, *139*, 273–284. [CrossRef]

27. Shu, W.; Hou, L.; Zhang, C.; Zhang, F.; Liu, J.; Liu, J.; Zhuang, L.; Zhang, J. Tailored Mg and Cu contents affecting the microstructures and mechanical properties of high-strength Al–Zn–Mg–Cu alloys. *Mater. Sci. Eng. A* **2016**, *657*, 269–283. [CrossRef]

28. Yang, X.; Chen, J.; Liu, J.; Qin, F.; Xie, J.; Wu, C. A high-strength AlZnMg alloy hardened by the T-phase precipitates. *J. Alloys Compd.* **2014**, *610*, 69–73. [CrossRef]

29. Zhao, Y.; Liao, X.; Jin, Z.; Valiev, R.; Zhu, Y.T. Microstructures and mechanical properties of ultrafine grained 7075 Al alloy processed by ECAP and their evolutions during annealing. *Acta Mater.* **2004**, *52*, 4589–4599. [CrossRef]

30. Li, Z.; Xiong, B.; Zhang, Y.; Zhu, B.; Wang, F.; Liu, H. Investigation of microstructural evolution and mechanical properties during two-step ageing treatment at 115 and 160 °C in an Al–Zn–Mg–Cu alloy pre-stretched thick plate. *Mater. Charact.* **2008**, *59*, 278–282. [CrossRef]

31. Hou, S.; Zhang, D.; Ding, Q.; Zhang, J.; Zhuang, L.-Z. Solute clustering and precipitation of Al-5.1Mg-0.15Cu-xZn alloy. *Mater. Sci. Eng. A* **2019**, *759*, 465–478. [CrossRef]

32. Berg, L.; Gjønnes, J.; Hansen, V.; Li, X.; Knutson-Wedel, M.; Waterloo, G.; Schryvers, D.; Wallenberg, L. GP-zones in Al–Zn–Mg alloys and their role in artificial aging. *Acta Mater.* **2001**, *49*, 3443–3451. [CrossRef]

33. Yang, X.; Liu, J.; Chen, J.; Wan, C.; Fang, L.; Liu, P.; Wu, C. Relationship Between the Strengthening Effect and the Morphology of Precipitates in Al-7.4Zn-1.7Mg-2.0Cu Alloy. *Acta Met. Sin. (Engl. Lett.)* **2014**, *27*, 1070–1077. [CrossRef]

34. Booth-Morrison, C.; Seidman, D.N.; Dunand, D.C. Effect of Er additions on ambient and high-temperature strength of precipitation-strengthened Al–Zr–Sc–Si alloys. *Acta Mater.* **2012**, *60*, 3643–3654. [CrossRef]

35. Michi, R.A.; Toinin, J.P.; Farkoosh, A.R.; Seidman, D.N.; Dunand, D.C. Effects of Zn and Cr additions on precipitation and creep behavior of a dilute Al–Zr–Er–Si alloy. *Acta Mater.* **2019**, *181*, 249–261. [CrossRef]
36. Zhang, J.; Wang, H.; Yi, D.; Wang, B.; Wang, H. Comparative study of Sc and Er addition on microstructure, mechanical properties, and electrical conductivity of Al-0.2Zr-based alloy cables. *Mater. Charact.* **2018**, *145*, 126–134. [CrossRef]
37. Yu, T.; Li, B.; Medjahed, A.; Hou, L.; Wu, R.; Zhang, J.; Sun, J.; Zhang, M. Impeding effect of the Al3(Er,Zr,Li) particles on planar slip and intergranular fracture mechanism of Al-3Li-1Cu-0.1Zr-X alloys. *Mater. Charact.* **2019**, *147*, 146–154. [CrossRef]
38. Zhang, C.; Yin, D.; Jiang, Y.; Wang, Y. Precipitation of L12-phase nano-particles in dilute Al-Er-Zr alloys from the first-principles. *Comput. Mater. Sci.* **2019**, *162*, 171–177. [CrossRef]
39. Bochvar, N.; Rybalchenko, O.; Leonova, N.; Tabachkova, N.; Rokhlin, L. Effect of cold plastic deformation and subsequent aging on the strength properties of Al-Mg2Si alloys with combined (Sc + Zr) and (Sc + Hf) additions. *J. Alloys Compd.* **2020**, *821*, 153426. [CrossRef]
40. Verdier, M.; Janecek, M.; Brechet, Y.; Guyot, P. Microstructural evolution during recovery in Al–2.5%Mg alloys. *Mater. Sci. Eng. A* **1998**, *248*, 187–197. [CrossRef]
41. Neubert, V.; Smola, B.; Stulíková, I.; Bakkar, A.; Reuter, J. Microstructure, mechanical properties and corrosion behaviour of dilute Al–Sc–Zr alloy prepared by powder metallurgy. *Mater. Sci. Eng. A* **2007**, *464*, 358–364. [CrossRef]
42. Li, H.; Liu, X.-T.; Wang, J.-Y. Influence of Preaging Treatment on Bake-Hardening Response and Electrochemical Corrosion Behavior of High Strength Al-Zn-Mg-Cu-Zr Alloy. *Metals* **2019**, *9*, 895. [CrossRef]
43. Bečvář, F.; Čížek, J.; Procházka, I.; Janotová, J. The asset of ultra-fast digitizers for positron-lifetime spectroscopy. *Nucl. Instrum. Methods Phys. Sect. A* **2005**, *539*, 372–385. [CrossRef]
44. Vlach, M.; Čížek, J.; Melikhova, O.; Stulíková, I.; Smola, B.; Kekule, T.; Kudrnová, H.; Gemma, R.; Neubert, V. Early Stages of Precipitation Process in Al-(Mn-)Sc-Zr Alloy Characterized by Positron Annihilation. *Met. Mater. Trans. A* **2015**, *46*, 1556–1564. [CrossRef]
45. Zhou, S.; Zhang, Z.; Li, M.; Pan, D.; Su, H.; Du, X.; Li, P.; Wu, Y. Effect of Sc on microstructure and mechanical properties of as-cast Al–Mg alloys. *Mater. Des.* **2016**, *90*, 1077–1084. [CrossRef]
46. Zhou, S.; Zhang, Z.; Li, M.; Pan, D.; Su, H.; Du, X.; Li, P.; Wu, Y. Formation of the eutecticum with multilayer structure during solidification in as-cast Al–Mg alloy containing a high level of Sc. *Mater. Lett.* **2016**, *164*, 19–22. [CrossRef]
47. Zhou, S.; Zhang, Z.; Li, M.; Pan, D.; Su, H.; Du, X.; Li, P.; Wu, Y. Correlative characterization of primary particles formed in as-cast Al-Mg alloy containing a high level of Sc. *Mater. Charact.* **2016**, *118*, 85–91. [CrossRef]
48. Li, J.; Wiessner, M.; Albu, M.; Wurster, S.; Sartory, B.; Hofer, F.; Schumacher, P. Correlative characterization of primary Al3(Sc,Zr) phase in an Al–Zn–Mg based alloy. *Mater. Charact.* **2015**, *102*, 62–70. [CrossRef]
49. Yang, Z.; Jiang, X.; Zhang, X.; Liu, M.; Liang, Z.; Leyvraz, D.; Banhart, J. Natural ageing clustering under different quenching conditions in an Al-Mg-Si alloy. *Scr. Mater.* **2021**, *190*, 179–182. [CrossRef]
50. Liu, S.; Li, C.; Han, S.; Deng, Y.; Zhang, X. Effect of natural aging on quench-induced inhomogeneity of microstructure and hardness in high strength 7055 aluminum alloy. *J. Alloys Compd.* **2015**, *625*, 34–43. [CrossRef]
51. Liu, M.; Čížek, J.; Chang, C.S.; Banhart, J. Early stages of solute clustering in an Al–Mg–Si alloy. *Acta Mater.* **2015**, *91*, 355–364. [CrossRef]
52. Chinh, N.Q.; Lendvai, J.; Ping, D.; Hono, K. The effect of Cu on mechanical and precipitation properties of Al–Zn–Mg alloys. *J. Alloys Compd.* **2004**, *378*, 52–60. [CrossRef]
53. Dellah, M.; Bournane, M.; Ragab, K.A.; Sadaoui, Y.; Sirenko, A. Early decomposition of supersaturated solid solutions of Al–Zn–Mg casting alloys. *Mater. Des.* **2013**, *50*, 606–612. [CrossRef]
54. Royset, J.; Ryum, N. Scandium in aluminium alloys. *Int. Mater. Rev.* **2005**, *50*, 19–44. [CrossRef]
55. Olafsson, P.; Sandström, R.; Karlsson, A. Comparison of experimental, calculated and observed values for electrical and thermal conductivity of aluminium alloys. *J. Mater. Sci.* **1997**, *32*, 4383–4390. [CrossRef]
56. Skoko, Ž.; Popović, S.; Štefanić, G. Microstructure of Al–Zn and Zn–Al Alloys. *Croat. Chem. Acta* **2009**, *82*, 405–420.
57. Khalfallah, A.; Raho, A.A.; Amzert, S.; Djemli, A. Precipitation kinetics of GP zones, metastable η′ phase and equilibrium η phase in Al−5.46wt.%Zn−1.67wt.%Mg alloy. *Trans. Nonferr. Met. Soc. China* **2019**, *29*, 233–241. [CrossRef]
58. Hou, S.; Liu, P.; Zhang, D.; Zhang, J.; Zhuang, L.-Z. Precipitation hardening behavior and microstructure evolution of Al–5.1 Mg–0.15Cu alloy with 3.0Zn (wt%) addition. *J. Mater. Sci.* **2017**, *53*, 3846–3861. [CrossRef]
59. Ganiev, I.N. High-Temperature and Electrochemical Corrosion of Aluminum-Scandium Alloys. *Protect. Met.* **1995**, *31*, 543–546.
60. Kharina, G.V.; Kochergin, V.P. Corrosion–Electrochemical Behavior of Al–Zn–rem Alloys in the Presence of Polyvanadate-ions. *Prot. Met.* **2001**, *37*, 575–579. [CrossRef]
61. Yoshioka, H.; Habazaki, H.; Kawashima, A.; Asami, K.; Hashimoto, K. The corrosion behavior of sputter-deposited Al Zr alloys in 1 M HCl solution. *Corros. Sci.* **1992**, *33*, 425–436. [CrossRef]
62. Shi, Y.; Pan, Q.; Li, M.; Huang, X.; Li, B. Effect of Sc and Zr additions on corrosion behaviour of Al–Zn–Mg–Cu alloys. *J. Alloys Compd.* **2014**, *612*, 42–50. [CrossRef]
63. Ghosh, K.; Gao, N.; Starink, M. Characterisation of high pressure torsion processed 7150 Al–Zn–Mg–Cu alloy. *Mater. Sci. Eng. A* **2012**, *552*, 164–171. [CrossRef]
64. Ghosh, K.; Gao, N. Determination of kinetic parameters from calorimetric study of solid state reactions in 7150 Al-Zn-Mg alloy. *Trans. Nonferr. Met. Soc. China* **2011**, *21*, 1199–1209. [CrossRef]
65. Belov, N.A.; Korotkova, N.; Akopyan, T.; Tsydenov, K. Simultaneous Increase of Electrical Conductivity and Hardness of Al–1.5 wt.% Mn Alloy by Addition of 1.5 wt.% Cu and 0.5 wt.% Zr. *Metals* **2019**, *9*, 1246. [CrossRef]

66. Belov, N.A.; Korotkova, N.O.; Akopyan, T.; Pesin, A. Phase composition and mechanical properties of Al–1.5%Cu–1.5%Mn–0.35%Zr(Fe,Si) wire alloy. *J. Alloys Compd.* **2019**, *782*, 735–746. [CrossRef]
67. Vlach, M.; Stulíková, I.; Smola, B.; Císařová, H.; Piešová, J.; Daniš, S.; Gemma, R.; Málek, J.; Tanprayoon, D.; Neubert, V. Phase transformations in non-isothermally annealed as-cast and cold-rolled AlMnScZr alloys. *Int. J. Mater. Res.* **2012**, *103*, 814–820. [CrossRef]
68. Myhr, O. Modelling of the age hardening behaviour of Al–Mg–Si alloys. *Acta Mater.* **2001**, *49*, 65–75. [CrossRef]
69. Myhr, O.R.; Grong, Ø.; Schäfer, C. An Extended Age-Hardening Model for Al-Mg-Si Alloys Incorporating the Room-Temperature Storage and Cold Deformation Process Stages. *Met. Mater. Trans. A* **2015**, *46*, 6018–6039. [CrossRef]
70. Starink, M. The determination of activation energy from linear heating rate experiments: A comparison of the accuracy of isoconversion methods. *Thermochim. Acta* **2003**, *404*, 163–176. [CrossRef]
71. Antonione, C.; Marino, F.; Riontino, G.; Abis, S.; Di Russo, E. An evaluation of AA 7012 microstructure by differential scanning calorimetry. *Mater. Chem. Phys.* **1988**, *20*, 13–25. [CrossRef]
72. Afify, N.; Gaber, A.-F.; Abbady, G. Fine Scale Precipitates in Al-Mg-Zn Alloys after Various Aging Temperatures. *Mater. Sci. Appl.* **2011**, *2*, 427–434. [CrossRef]
73. Abis, S.; Riontino, G. A resistivity study of 7012 Al-Zn-Mg alloy commercial tempers. *Mater. Lett.* **1987**, *5*, 442–445. [CrossRef]
74. Fan, X.; Jiang, D.; Meng, Q.; Zhong, L. The microstructural evolution of an Al–Zn–Mg–Cu alloy during homogenization. *Mater. Lett.* **2006**, *60*, 1475–1479. [CrossRef]
75. Styles, M.; Hutchinson, C.; Chen, Y.; Deschamps, A.; Bastow, T. The coexistence of two S (Al2CuMg) phases in Al–Cu–Mg alloys. *Acta Mater.* **2012**, *60*, 6940–6951. [CrossRef]

 *metals*

Article

Ageing of Al-Mn-Cu-Be Alloys for Stimulating Precipitation of Icosahedral Quasicrystals

Tonica Bončina [1],*, Mihaela Albu [2] and Franc Zupanič [1]

[1]   Faculty of Mechanical Engineering, University of Maribor, SI-2000 Maribor, Slovenia; franc.zupanic@um.si
[2]   Graz Centre for Electron Microscopy, A-8010 Graz, Austria; mihaela.albu@felmi-zfe.at
*   Correspondence: tonica.boncina@um.si; Tel.: +386-2-220-7863

Received: 27 May 2020; Accepted: 9 July 2020; Published: 11 July 2020

Abstract: In this work, the ageing of some Al-Mn-Cu-Be alloys was investigated in the temperature range in which predominantly icosahedral quasicrystalline (IQC) precipitates can form. The alloys were cast into a copper mould, directly aged (T5 heat treatment) between 300 and 440 °C for different times. Afterwards, they were examined using scanning and transmission electron microscopy, X-ray diffraction and hardness testing. The main aim of the work was to determine the conditions at which a high number density of spherical icosahedral quasicrystalline precipitates can form. The highest number density of IQC precipitates was obtained at 300 °C after prolonged ageing. The spheroidal precipitates had a diameter less than 20 nm. The size of IQC precipitates increased with the increasing temperature, and in addition, decagonal quasicrystalline precipitates appeared. The time to maximum hardness decreased strongly with increasing ageing temperature. The IQC precipitates can form in a fairly broad temperature range in Al-Mn-Cu-Be alloys and that by varying ageing temperature and duration, rather different distributions of precipitates can be obtained. The presence of precipitates caused rather strong aluminium alloys and fast work hardening during initial plastic deformation.

Keywords: aluminium; ageing; quasicrystal; transmission electron microscopy

1. Introduction

Quasicrystals were discovered in Al-Mn alloys [1]. Since an icosahedral quasicrystalline phase (IQC) in these and similar alloys is metastable [2], the alloys have been predominantly fabricated by rapid solidification and then, compacted [3]. Rapid solidification enables the formation of very small particles, which can slightly coarsen during compaction, but a fine distribution of quasicrystalline particles can provide high strength [4]. The alloying of Al-Mn alloys with Be, Ce, Si or Fe can stimulate the formation of IQC particles during solidification with slower cooling rates [5–8], however, their sizes are coarser, which decreases their strength and toughness.

A fine distribution of particles can be obtained by precipitation hardening, which is a typical hardening process in many aluminium alloys [9–12]. There are several types of precipitates, which are mainly metastable. According to the best of the authors' knowledge, there are no commercial alloys that would be mainly strengthened by icosahedral quasicrystalline particles. The manufacturing of alloys with a high density of IQC precipitates may be a challenge and they can also have an interesting combination of properties. The formation of IQC precipitates was observed in many Al-alloys. Hansen and Gjonnes [13] found that the IQC phase formed during heat treatment of a high-Mn aluminium alloy, and then, transformed to the cubic α-AlMnSi phase. Li and Arnberg [14] studied dispersoids in a commercial 3003 alloy. Some icosahedral quasicrystal dispersoids were found to precipitate at the early stage of precipitation. Mochugovskiy et al. [15] discovered IQC precipitates in an Al-3%Mg-1%Mn alloy during annealing at 360 °C. Partly coherent IQC precipitates predominantly formed on the dislocations and dislocation walls, and were relatively rare.

The IQC precipitates were discovered also in an $Al_{94}Mn_2Be_2Cu_2$ alloy during in situ study of the temperature stability of the IQC phase formed during solidification [16]. During further research, they were found in samples heat-treated at 300 and 400 °C [17]. They did not form at 200 °C because of the very low diffusivity of Mn. Instead, some precipitation of Θ'-Al_2Cu was observed. The T-$Al_{20}Mn_3Cu_2$ precipitates formed at 500 °C, which are also known from other alloys containing Cu and Mn [18]. The IQC precipitates in this alloy have primitive icosahedral structure (space group $Pm\overline{3}5$). They form at least a semi-coherent interface with an aluminium matrix, and possess a specific orientation relationship with it. All aluminium fourfold axes are parallel to three twofold axes of IQC, and at the same time also, threefold aluminium axes are parallel with the threefold axes of IQC [17,19].

Since the precipitation of quasicrystalline precipitates takes place within a temperature range, in which most aluminium alloys rapidly lose their strength, the alloys with quasicrystalline precipitates may be used at high temperature application, e.g., produced by high pressure die-casting or single roll casting, which have moderate cooling rates. One of the main prerequisites is the presence of sufficiently high volume fraction of uniformly distributed IQC precipitates, which has not been achieved in Al-alloys yet.

In our previous work, it was found out that icosahedral quasicrystalline precipitates can form in Al-Mn-Cu-Be alloy during T5 heat treatment at 300 and 400 °C [17]. The IQC precipitate possessed good matching with the Al-rich solid solution α-Al, and have a specific orientation relationship with it [17,19]. It was also discovered that in the alloy Al-Mn-Cu-Be-Sc-Zr, plate-like decagonal quasicrystals formed on the {001} planes of α-Al [19]. These suggested that the Al-Mn-Cu-Be alloys could be used for precipitation hardening. In order to prove this, two Al-Mn-Cu-Be alloys with improved chemical compositions were cast into copper moulds with different diameters to achieve cooling rates 250–600 °C/s and then, aged in the temperature range between 300 and 440 °C for different durations. The primary aim of this investigation was to find the conditions at which a high density of spherical icosahedral precipitates in an Al matrix can be obtained by T5 heat treatment (ageing of samples in the as-cast condition). In addition, the properties of the stronger alloy at room temperature were determined, as well as the microstructures after some degree of plastic deformation.

2. Materials and Methods

The compositions of the investigated alloys are given in Table 1. The manganese content is essential for the formation of quasicrystalline phase during solidification. The addition of beryllium enables formation of the quasicrystalline phase at lower cooling rates and also stimulates formation of quasicrystalline precipitates. By the addition of copper, T-$Al_{20}Mn_3Cu_2$ precipitates can form at higher temperatures, and by application of T6 heat treatment, Cu-rich precipitates can be created. The contents of alloying elements was decreased comparing to the previously studied Al-Mn-Cu-Be alloy, in which IQC precipitates were determined [17].

Table 1. Chemical compositions of the investigated alloys after vacuum induction melting, as determined by Atomic Emission Spectroscopy–Inductively Coupled Plasma (w—weight fraction in %, x—atomic fraction in %).

Alloy No.	Alloy	w (Al)	w (Mn)	w (Be)	w (Cu)	x (Al)	x (Mn)	x (Be)	x (Cu)
#1	$Al_{96.5}Mn_{0.8}Cu_{1.2}Be_{1.5}$	Remain	1.55	0.48	2.73	Remain	0.77	1.47	1.18
#2	$Al_{95.4}Mn_{1.8}Cu_{1.6}Be_{1.2}$	Remain	3.51	0.38	3.71	Remain	1.78	1.18	1.63

The alloys were vacuum induction melted using commercially available Al99.8, AlMn10, AlCu25 and AlBe5.5 master alloys into bars with diameters 20 mm, and sectioned to pieces of about 200 g. The pieces were melted and cast into a copper block with a diameter of 110 mm and height of 120 mm, with the casting temperature around 750 °C. The melt solidified in cylindrical holes engraved into the copper block with the diameters of 4, 6 and 10 mm and lengths of 40 mm. The solidification rates before beginning solidification, determined experimentally, were 580 ± 53, 510 ± 44 and 258 ± 31 K s^{-1}

for 4, 6 and 10 mm rod, respectively. After casting, the cast samples were aged artificially in the air at temperatures from 300 to 440 °C for different times. This kind of heat treatment is designated as T5 treatment. During T6 teat treatment, T-$Al_{20}Mn_3Cu_2$ precipitates form on solution annealing, and almost no alloying elements remain available for the formation of quasicrystalline precipitates during ageing.

The basic metallographic analysis was done by Light Microscopy (Epiphot 300, Nikon, Tokyo, Japan) and Scanning Electron Microscopy (Sirion 400 NC, FEI, Eindhoven, The Netherlands) equipped with an Energy Dispersive Spectrometer (INCA x-sight, Oxford Analytical, Bicester, UK). Some samples were investigated after final polishing with a diamond paste (1 μm) only. The samples were etched with a methanol-iodine solution after ageing, which revealed the distribution of precipitates in the microstructure and precipitation free zones [20]. The phase analysis was carried out by XRD (X-ray Diffraction, Sincrotrone Elettra, Trieste, Italy) using synchrotron X-rays with a wavelength of 99.9996 pm. Lamellas for the Transmission Electron Microscopy (TEM) were prepared using a Focused Ion Beam FIB (Helios, FEI, Eindhoven, The Netherlands). High-Resolution TEM (Titan[3] G2 60–300, FEI, Eindhoven, The Netherlands) and Energy-Dispersive X-ray Spectroscopy, EDS (SuperX, Bruker, Billerica, MA, USA) were carried out in an probe corrected electron microscope. The compression test was carried out by a quenching/deformation dilatometer (DIL 805A/D, TA Instruments, New Castle, DE, USA) at room temperature, with the strain rate 0.05 s^{-1}. The sample had 10 mm length and diameter 5 mm. It was machined from the cast bar with a diameter of 6 mm. Two samples were deformed to 0.09 and 0.4 true strain and investigated by TEM. Hardness was measured by a Zwick 3212 with a load of 0.1 kg (HV 1).

3. Results

3.1. As-Cast Condition

In samples with diameters 4 and 6 mm, the columnar α-Al grains with a cellular-dendritic morphology prevailed. Some equiaxed grains were at the centres of the 10 mm in diameter samples. The microstructures of all alloys in the as-cast condition consisted predominantly of three phases, α-Al (Al-rich solid solution), icosahedral quasicrystal (IQC) and Θ-Al_2Cu (Figure 1).

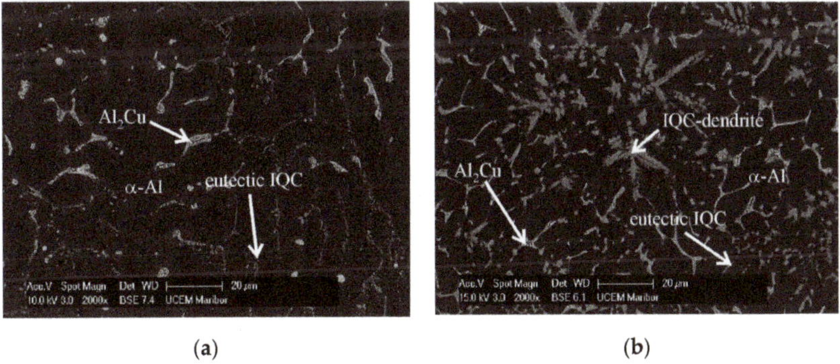

(a)   (b)

Figure 1. Backscattered electron micrographs of the gravitationally cast samples in the as-cast conditions of (**a**) the alloy $Al_{96.5}Mn_{0.8}Cu_{1.2}Be_{1.5}$, (**b**) $Al_{95.4}Mn_{1.8}Cu_{1.6}Be_{1.2}$. Polished samples.

The primary dendritic IQC phase, with dendrite arms extending in the threefold directions, was only present at the centre of the 10 mm sample in the alloy $Al_{95.4}Mn_{1.8}Cu_{1.6}Be_{1.2}$ (Figure 1b). The IQC phase was present as a part of a (α-Al + IQC) binary eutectic (Figure 1). The Θ-Al_2Cu phase was in the interdendritic spaces as a heterogeneous constituent (α-Al + Θ-Al_2Cu). The dendrite arm spacing and the interdendritic areas increased with the specimen size.

The XRD revealed the same phases in the as-cast state as SEM micrographs, α-Al, IQC and Θ-Al$_2$Cu (Figure 2a). The lattice parameters of the FCC α-Al were similar in all alloys and samples. They were about 403.80 pm, which is 0.28% less than in pure Al (powder diffraction file pdf 000-04-0787). This value indicates that the matrix was supersaturated with Cu and Mn, having the smaller atomic diameters than Al (R_{Al} = 143.2 pm, R_{Mn} = 136.7 pm, R_{Cu} 127.8 = pm). The calculated lattice parameter of a(Al) would be 403.82 pm if we apply Vegard's law and assume the maximum equilibrium solubility of Cu and Mn in Al (Mn: 0.62 at. % [21] and Cu: 2.5 at. % [22]). Almost the same values of measured and calculated a(Al) strongly suggest a high supersaturation of the matrix. The lattice parameters of Θ-Al$_2$Cu were nearly the same as in pure Θ-Al$_2$Cu (pdf 000-89-198), pointing out that it did not dissolve Mn or Be. The IQC phase exhibited wide peaks, arising from small particle sizes and the presence of many defects in them. A strong peak at approximately 27.5° occurred in a 10 mm sample of the alloy Al$_{95.4}$Mn$_{1.8}$Cu$_{1.6}$Be$_{1.2}$, which is designated by (!) in Figure 2a. Its presence testified the existence of an additional phase at the sample centre, belonging probably to Al$_{15}$Mn$_3$Be$_2$ [6] or any other metastable phases found in Al-Mn-Cu alloys by Stan-Glowinska [23].

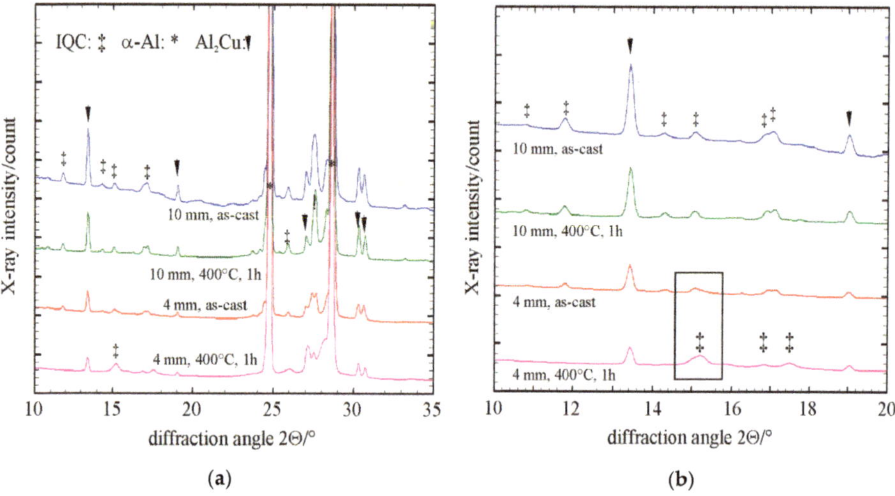

Figure 2. X-ray diffraction patterns of different samples and conditions of the Al$_{95.4}$Mn$_{1.8}$Cu$_{1.6}$Be$_{1.2}$ alloy. (**a**) The range between 10° and 35°, and (**b**) the range between 10° and 20° that shows clearly the positions and shapes of minor phases' peaks.

The highly alloyed Al$_{95.4}$Mn$_{1.8}$Cu$_{1.6}$Be$_{1.2}$ attained the highest hardness in the as-cast state due to the higher phase fraction of IQC and Θ-Al$_2$Cu (Figure 3a). The hardness of smaller specimens was much higher due to their finer microstructure with shorter secondary arm spacing, smaller interdendritic spaces and finer IQC and Θ-Al$_2$Cu particles (Figure 3b).

3.2. Heat Treatment

Figure 3 shows the ageing behaviour of the alloys after being heated from the as-cast state to the ageing temperatures of 300 and 400 °C. The hardness dropped immediately after the beginning of ageing in all cases. The hardness recovery started after several hours at 300 °C, and about after half an hour at 300 °C. For the most alloys, the hardness attained the highest value approximately after one week of ageing (168 h) at 300 °C and thereafter, did not change a lot. The maximum hardness was attained after 1 h of ageing at 400 °C. Thereafter, only a slow decrease of hardness was noticed. The hardness of the smaller samples (4 and 6 mm) was much higher than the 10 mm sample. The properties did not fall below the hardness in the as-cast condition, even after the longest ageing times.

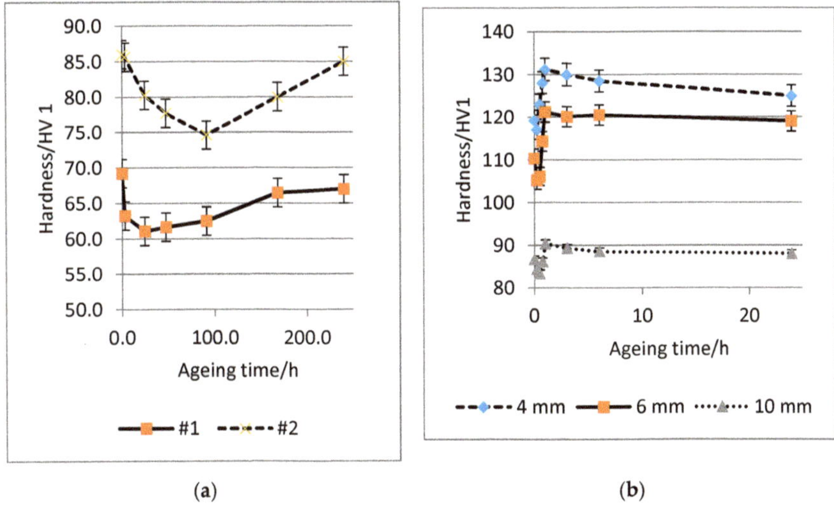

Figure 3. Hardness of the alloys in the as-cast state and after heat treatment. (**a**) Hardness of different alloys aged at 300 °C (sample diameter 10 mm), (**b**) hardness of different samples of the alloy $Al_{95.4}Mn_{1.8}Cu_{1.6}Be_{1.2}$ heat-treated at 400 °C. (#1—$Al_{96.5}Mn_{0.8}Cu_{1.2}Be_{1.5}$, #2—$Al_{95.4}Mn_{1.8}Cu_{1.6}Be_{1.2}$).

Figure 4 shows the backscattered electron micrographs of some alloys aged at 300 and 400 °C. The primary and eutectic IQC particles hardly changed, while Θ-Al_2Cu became more compact. The micrographs at higher magnifications clearly showed that there were many precipitates in the α-Al matrix. Their sizes were smaller than 100 nm. The images indicate that the particles have mainly spherical shape, but it was difficult to discern more details from the SEM micrographs.

Heat treatment did not considerably change the XRD diffractograms (Figure 2). The aluminium peaks moved slightly to smaller angles because the lattice parameters of α-Al increased to 404.48 ± 0.03 pm for all alloys and samples aged at 300 and 400 °C. The supersaturation with Cu and Mn considerably decreased along with precipitation. The fraction of Θ-Al_2Cu slightly decreased, while its lattice parameters did not change. The positions and shapes of peaks in the 10 mm sample of the alloy $Al_{95.4}Mn_{1.8}Cu_{1.6}Be_{1.2}$ remain almost the same, suggesting that the structure of primary and IQC did not change. On the other hand, the peak at 15.20° became wider and moved to a higher angle (indicated by the rectangle). This is a result of precipitation of IQC in the matrix that also has a smaller lattice parameter than primary and eutectic IQC. Its value was about 0.46 [24], the same as in binary Al-Mn alloys [25]. However, the IQC precipitates had a quasilattice constant of 0.35 nm in an Al-Mn-Cu-Be-Sc-Zr alloy [19], and the same happened in this case.

For the TEM investigation, the 4 mm diameter samples of the alloy $Al_{95.4}Mn_{1.8}Cu_{1.6}Be_{1.2}$ were chosen because they had the most uniform and finest microstructure. In addition, the hardness of these samples was higher, indicating the highest values that can be achieved by heat treatment of castings.

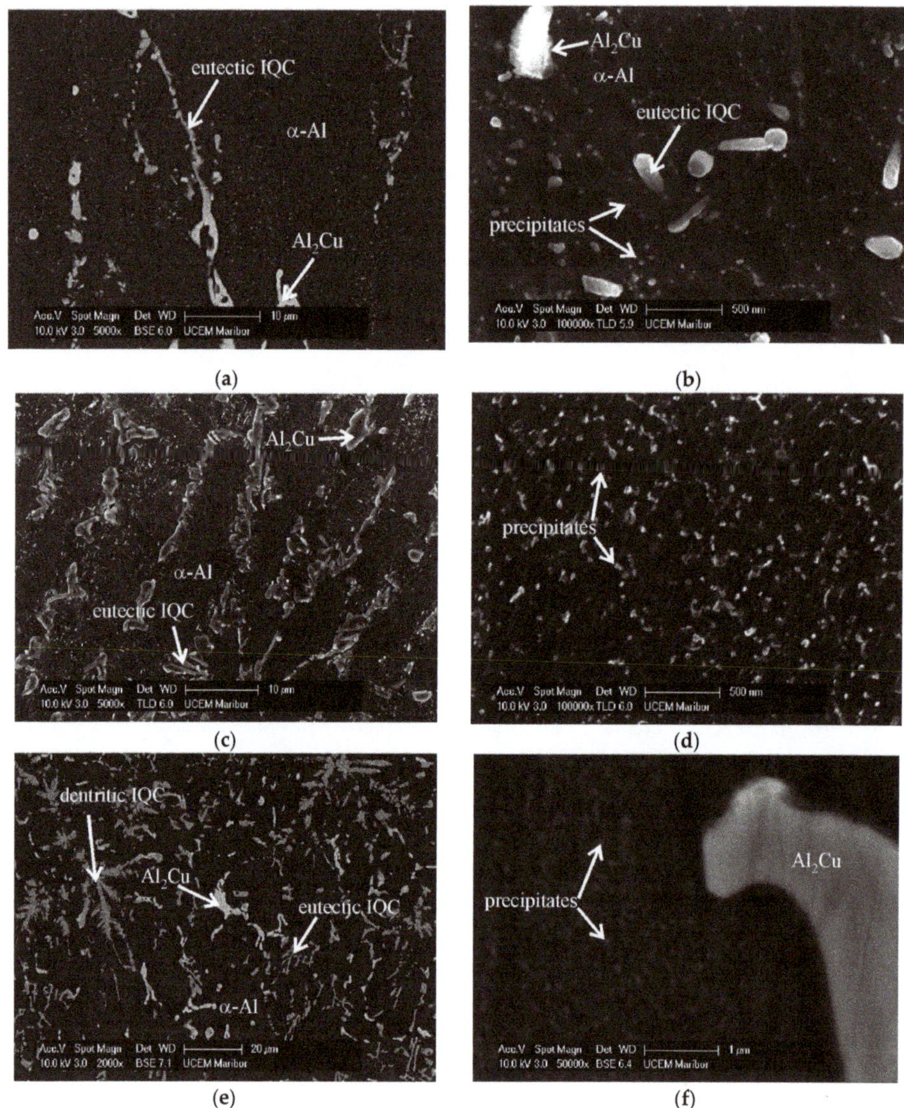

(a) (b) (c) (d) (e) (f)

Figure 4. Backscattered electron micrographs of the alloys. (**a,b**) $Al_{96.5}Mn_{0.8}Cu_{1.2}Be_{1.5}$ and (**c,d**) $Al_{95.4}Mn_{1.8}Cu_{1.6}Be_{1.2}$ aged at 300 °C for 168 h, and (**e,f**) the alloy $Al_{95.4}Mn_{1.8}Cu_{1.6}Be_{1.2}$ aged at 400 °C for 1 h. The diameter of samples was 10 mm. The samples were etched using a methanol-iodine solution.

Figure 5a shows the typical microstructure of a 4 mm sample. The cell diameters are 3–5 µm, which is 2–3 times smaller than in the 10 mm samples. IQC and Θ-Al_2Cu were at the cell boundaries (Figure 5b,c). The precipitation free zone was around cell boundaries, and there was a high precipitate density within the cells. Figure 6 shows precipitates in samples aged at 300 and 400 °C. At 300 °C, a very high density of spheroidal precipitates formed that were smaller than 20 nm (16.9 ± 1.5 nm). At some places, even three precipitates were one above another within the TEM lamella, which was about 80–90 nm thick. In such cases, the distances between particle centres could be about 30 nm

(Figure 6b). The interparticle distances were up to 50 nm elsewhere. At 400 °C, slightly larger spheroidal particles were formed (26.4 ± 2.5 nm), accompanied by plate-like particles with the thickness of 15.6 ± 3.0 nm, and the length 58.4 ± 7.6 nm. The interparticle distances were larger than at 300 °C. No other precipitates (e.g., Cu-rich (Θ-Al$_2$Cu, θ' or θ''), T-Al$_{20}$Mn$_3$Cu or Be-rich precipitates) were identified in the microstructure.

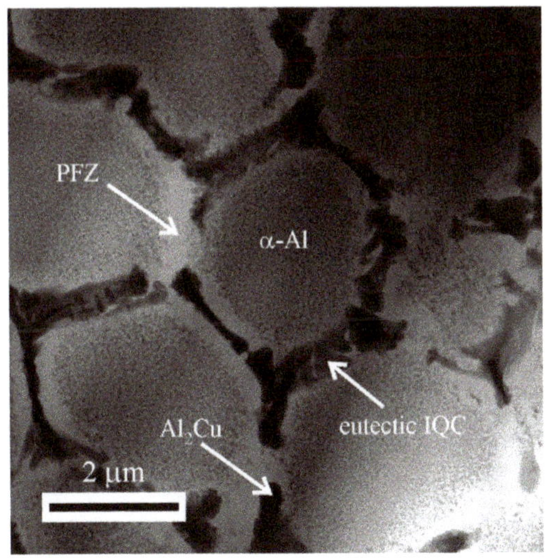

(a)

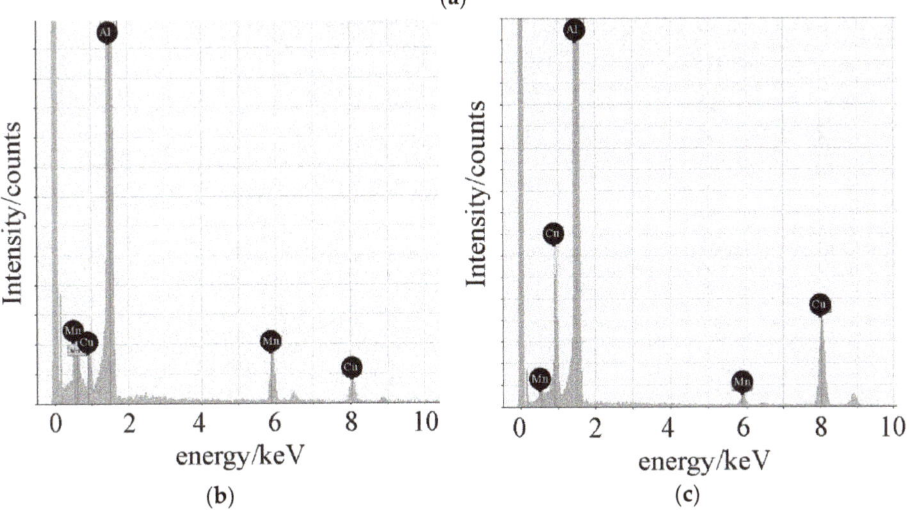

(b)   (c)

Figure 5. (**a**) Bright field transmission electron micrograph of the alloy Al$_{95.4}$Mn$_{1.8}$Cu$_{1.6}$Be$_{1.2}$ after ageing at 400 °C for 1 h; sample diameter 4 mm (PFZ—precipitation free zone). (**b**) EDS spectrum of the eutectic IQC and (**c**) of Θ-Al$_2$Cu.

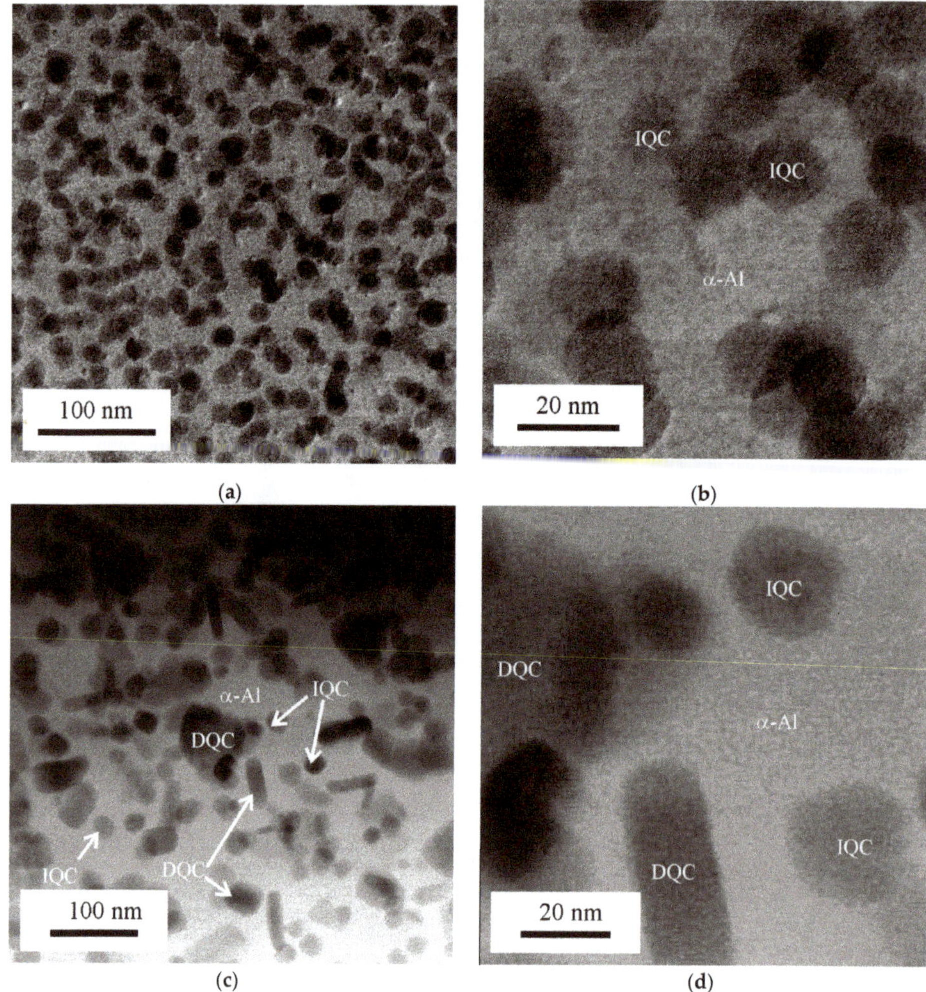

Figure 6. Bright-field TEM micrographs of the alloy $Al_{95.4}Mn_{1.8}Cu_{1.6}Be_{1.2}$ (**a,b**) heat-treated at 300 °C for 168 h and (**c,d**) heat-treated at 400 °C for 1 h (orientation close to [001]-Al). Sample diameter 4 mm. (IQC—icosahedral quasicrystal, DQC—decagonal quasicrystal).

3.3. Analysis of Icosahedral and Decagonal Quasicrystalline Precipitates

Figure 7a shows a HRTEM image of a spheroidal precipitate in α-Al. The EDS in TEM showed that it consisted of Al, Mn and some Cu (Figure 7b). EDS was not able to reveal Be. The fast-Fourier-transform (FFT) of the whole area showed the superposition of two patterns, which are shown separately in Figure 7d,e. The distances between the peaks in the directions denoted by arrows were not the same, and increased with the golden mean τ (Figure 7c,e). This clearly showed the quasicrystalline nature of the spherical precipitates; their icosahedral quasicrystalline structure was identified in detail in the previous works [16,17,19].

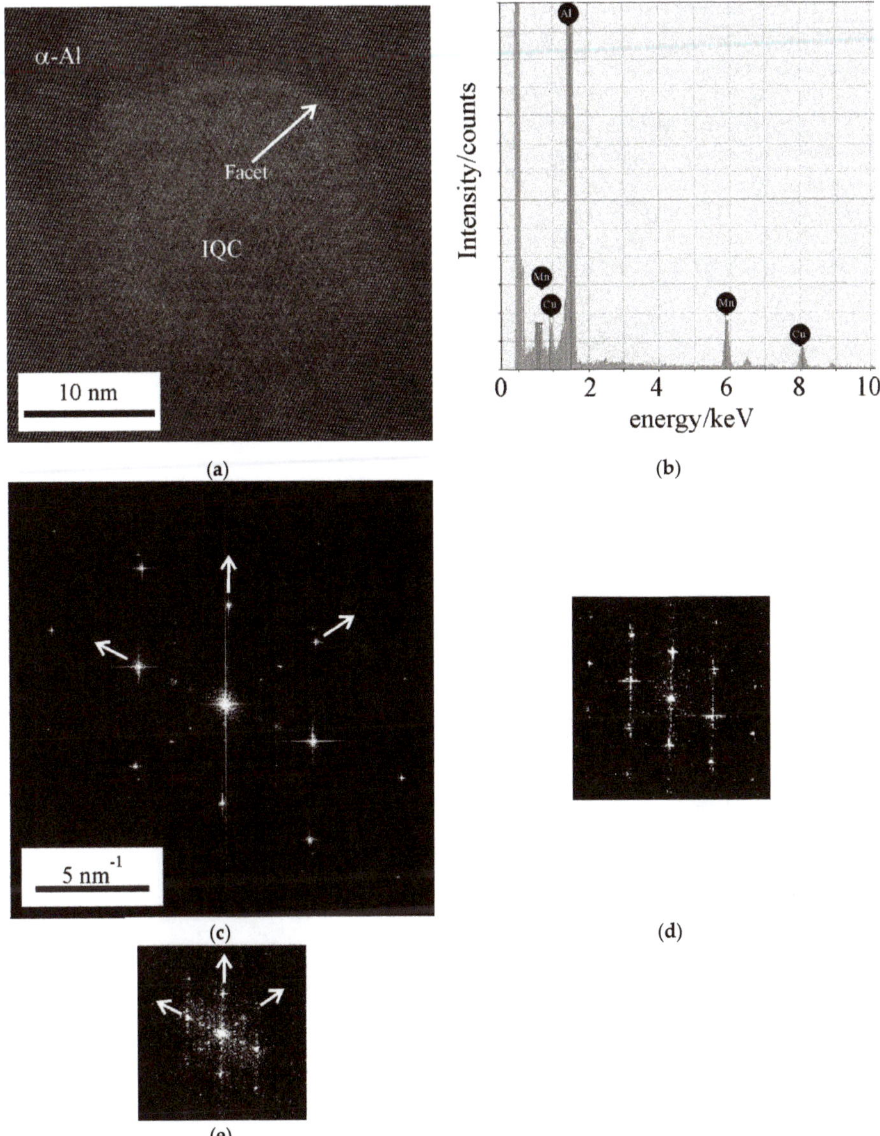

Figure 7. An icosahedral quasicrystalline (IQC) precipitate in $Al_{95.4}Mn_{1.8}Cu_{1.6}Be_{1.2}$ after ageing at 400 °C for 1 h. (**a**) A high-resolution scanning transmission electron image, (**b**) an EDS spectrum of the IQC particle, (**c**) Fast Fourier transform (FFT) of all area, (**d**) FFT of Al-rich matrix, with the zone axis [111], (**e**) FFT of the IQC particle, with the threefold zone axis $[011 0\bar{1}0]$.

A closer inspection showed that the threefold axis of the α-Al [111] was parallel to the threefold axis of the IQC $[011 0\bar{1}0]$ (indices according to Singh and Ranganathan [26]). The same orientation relationship between α-Al and IQC was found in a melt-spun Al-Mn-Si alloy [27]. A quasilattice constant of 0.35 nm was inferred, which is smaller than in the Al-Mn quasicrystal, calculated using the Elser's scheme [25]. HRTEM reveals that the IQC precipitate was not completely spherical but showed tendency to faceting. This was also found by IQC particles formed during rapid solidification [28,29].

IQC quasicrystals have predominantly shapes of pentagonal dodecahedrons, in which each facet has a shape of a regular pentagon. Due to a high symmetry of icosahedral quasicrystal, the shape appears spherical at lower magnifications.

The plate-like precipitates were identified as decagonal quasicrystalline precipitates (Figure 8a). They also consisted mainly of Al, Mn, with a small amount of Cu (Figure 8b). Their predominant habitus planes represented all three {001} planes of the α-Al. The FFTs of the HRTEM image showed the orientation relationship between DQC and FCC matrix (Figure 8c–e). For a precipitate lying on the (001) plane of the α-Al, its tenfold axis was parallel to the [010] direction of the α-Al, while its twofold axis was parallel to the [100] direction of the α-Al. The same orientation relationship was also found in the Al-Mn-Cu-Be alloys, containing Sc and Zr [19].

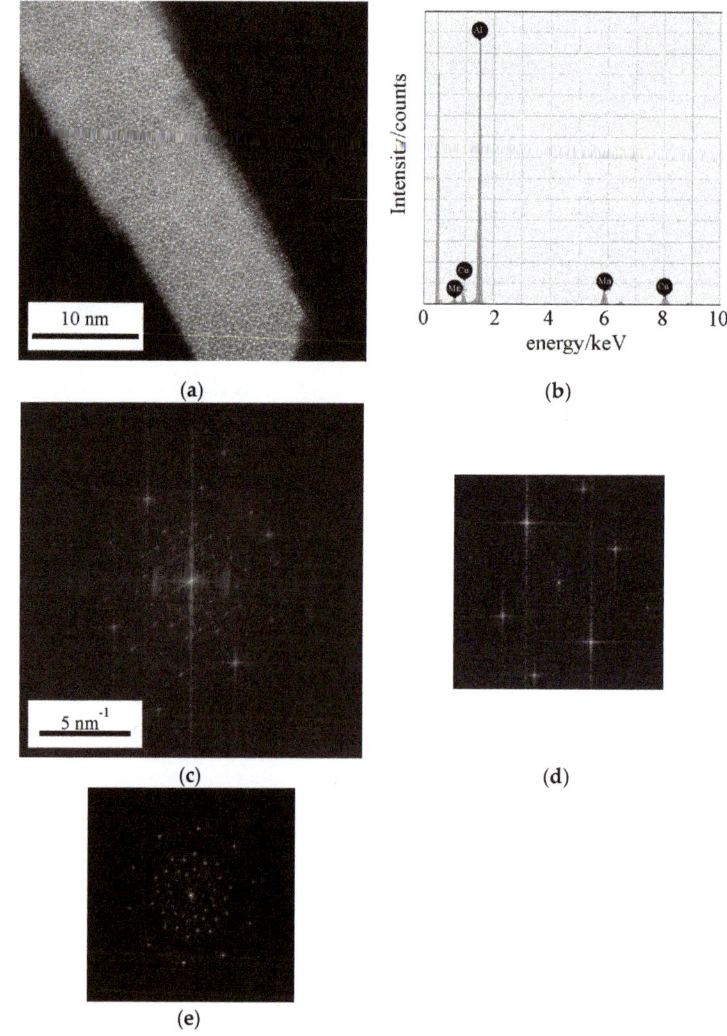

Figure 8. A decagonal quasicrystalline (DQC) precipitate in $Al_{95.4}Mn_{1.8}Cu_{1.6}Be_{1.2}$ after ageing at 400 °C for 1 h. (**a**) A high-resolution scanning transmission electron image, taken with a HAADF detector (High-angle Annular Dark Field), (**b**) an EDS spectrum of the IQC particle, (**c**) FFT of the entire area, (**d**) FFT of Al-rich matrix, with the zone axis [001], (**e**) FFT of the DQC particle along its tenfold axis.

3.4. Compression Test and Microstructures of Deformed Samples

Figure 9 shows the stress–strain diagram of the alloy $Al_{95.4}Mn_{1.8}Cu_{1.6}Be_{1.2}$ at room temperature. The yield stress σ_y is slightly above 300 MPa. At the beginning, the work hardening is very high, and the alloy attained the highest true stress σ_m = 461 MPa at true strain φ = 0.1. The softening took place up to $\varphi \approx 0.6$. The final hardening occurred until the appearance of cracks on the specimen surface at $\varphi \approx 0.8$. The barrelling started to occur at a deformation of 0.4 but was rather slight (insert in Figure 9). Thus, the value of the true stress at $\varphi \approx 0.8$ should be up to 5% lower.

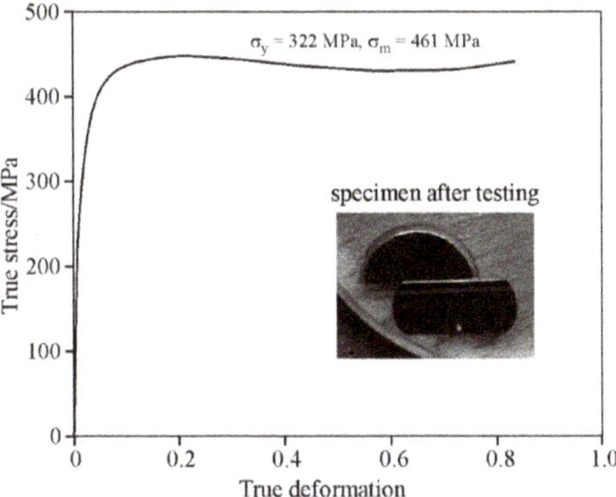

Figure 9. The compression test of the alloy $Al_{95.4}Mn_{1.8}Cu_{1.6}Be_{1.2}$ after ageing at 400 °C for 1 h, 6 mm sample. The specimen after testing is shown.

Figure 10 shows TEM micrographs after compression for φ = 0.1, which is close to the maximum on the compression curve. This deformation caused fracture of IQC and Θ-Al_2Cu at cell centres (Figure 10a). Darker regions in Figure 11b have higher dislocation density. The dislocations were almost absent in PFZ and areas with lower precipitate density (Figure 10c), but much higher in the areas with higher precipitate densities. In spite of good matching with the matrix, it is rather unlikely that dislocations could cut the particles (Figure 10d). It is more probable that the Orowan mechanism is responsible for the strengthening.

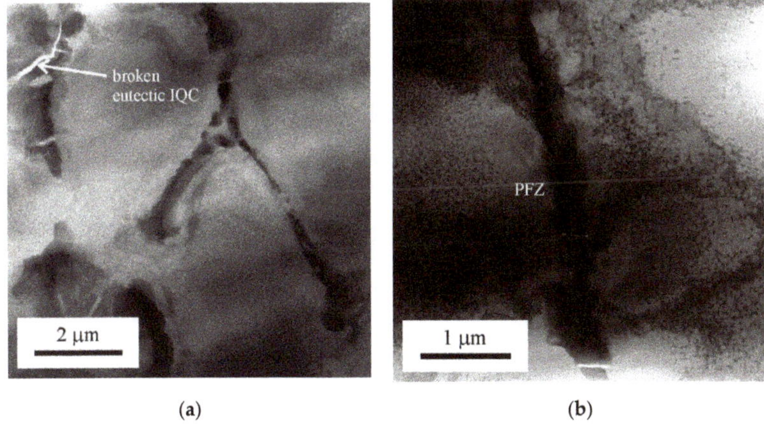

Figure 10. *Cont.*

(c) (d)

Figure 10. Bright field TEM micrographs of the alloy $Al_{95.4}Mn_{1.8}Cu_{1.6}Be_{1.2}$ aged at 400 °C for 1 h and compressed with 0.09 true stress. (**a**) Deformed cells, (**b**) cell boundary with the precipitation free zone, (**c**) areas with different precipitate densities, (**d**) an area, with a higher precipitate density.

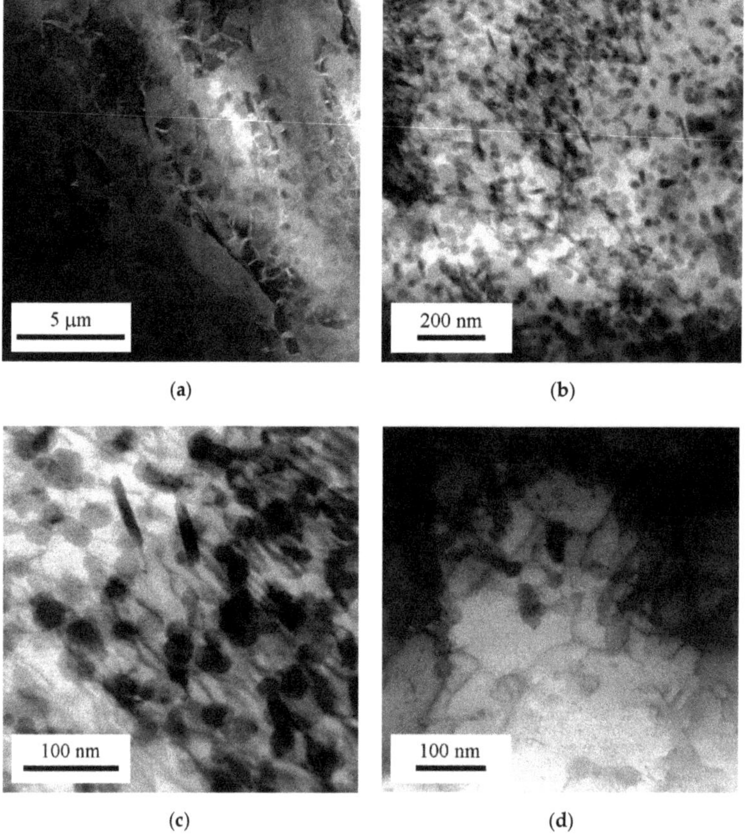

(a) (b)

(c) (d)

Figure 11. Bright field TEM micrographs of the alloy $Al_{95.4}Mn_{1.8}Cu_{1.6}Be_{1.2}$ aged at 400 °C for 1 h and compressed with 0.4 true strain. (**a**) Broken phases at a cell boundary, (**b**) an area inside a cell, (**c**) an area, with a higher precipitate density, (**d**) an area, with a lower precipitate density.

Figure 11 depicts TEM micrographs after compression for $\varphi = 0.4$. This deformation caused fragmentation of IQC and Θ-Al$_2$Cu formed during solidification (Figure 11a). The dislocations were more uniformly distributed (Figure 11b), however, there was still higher density of dislocations in regions with a higher precipitation density (Figure 11c,d).

4. Discussion

The supersaturation of the solid solution matrix with alloying elements and vacancies is a prerequisite for the formation of precipitates during ageing [30]. By T5 treatment, an alloy is cooled from an elevated temperature shaping process and then, artificially aged. In this investigation, casting represented the elevated shaping process and supersaturation can be achieved by sufficiently fast cooling from the liquid state, which can cause the fine microstructure besides supersaturation. The examination revealed the formation of a metastable IQC phase during solidification. The samples of all alloys with diameters from 4 to 10 mm have the same lattice parameters in the as-cast state, indicating similar supersaturation with Cu and Mn, which was close to their maximum equilibrium solubility in α-Al. This suggests the similar attainable volume fraction of precipitates in all alloys. Further increase of supersaturation can be achieved by rapid solidification, but not by typical casting processes. The Be content is essential for the formation of the quasicrystalline phase during solidification with the cooling rates obtained in the copper mould (250–600 K s^{-1}). It is incorporated in the IQC phase in rather high amounts [24,31].

The equilibrium intermetallic phases at ageing temperatures of the investigated alloys are expected to be Θ-Al$_2$Cu and τ_1 from the ternary Al-Mn-Cu system [32], while Be should be found in Be$_4$Al(Cu,Mn) [24]. None of these phases appeared during ageing. The microstructure hardly changed at the ageing temperatures, indicating that only a small amount of primary phases dissolved. Thus, only the elements already present in the solid solution could contribute to the precipitation. The ageing temperature was above the solvus lines for GP zones, Θ'' and Θ' Cu-rich precipitates [22]. There were no rod-like T-Al$_{20}$Mn$_3$Cu$_2$ that could be as well formed in Al-Mn-Cu-Be alloys at higher temperatures and in many other Al-alloys containing Cu and Mn [17,33].

The dominant phase that occurred in the studied temperature range was IQC. A high-resolution electron image (Figure 7a) showed excellent matching with α-Al, which was found also in other investigations [16,17,19]. Good matching between quasicrystalline and periodic crystals was found in many systems by Singh and Tsai [34–36], which is based on a high symmetry of IQC that can allow good correspondence between different planes of IQC and periodic crystals. Good matching also lowers the interfacial energy. In this alloy, the faceting tendency of IQC precipitates with a diameter of only 25 nm was observed. It is likely that they adopt a shape of pentagonal dodecahedron, which is surrounded by high density pentagons, further decreasing the surface energy. The spheroidal shape of IQC precipitates also implies low deformation energy by their formation. IQC precipitates have been found in many Al-alloys [14,15,37], however, their number density was rather low. It is believed that Be plays an important role by the formation of IQC precipitates. Beryllium was found at the centre of the IQC phase formed during the solidification, because it stimulates the icosahedral order in the melt [31]. It seems highly probable that Be acts in an analogous way in the solid state and initiates the formation of the quasicrystalline precipitates. Its atomic diameter ($R = 111.3$ pm) is about 22% less than Al. When a Be atom substitutes an Al atom at the lattice site, it can cause considerable distortion of the lattice, which may stimulate the nucleation of an IQC precipitate. After nucleation, the growth of the spheroidal IQC precipitates is limited by the diffusion of Mn, which has a much smaller diffusion coefficient than copper [38]. The growth of IQC precipitates is rather slow, since the coarsening rate of spherical precipitates is proportional to the square root of time [39].

In Al-Mn-Cu alloys, decagonal quasicrystals can form upon solidification [23,40]. The decagonal Al-Mn-Cu phase is closely related to the equilibrium phase τ_1. Thus, it is surely more thermodynamically stable than IQC. However, it is much less symmetrical. It causes more deformation during formation, thus, it appeared in a plate-like shape. Its habitus planes are (001), because they have the lowest modulus of elasticity (62 GPa compared to 76 GPa of (111) planes). The growth of decagonal plates can take place linearly with time, so their lengthening rate is much higher than that of IQC precipitates. As a result, the appearance of DQC precipitates is a disadvantage, when considering the number density of precipitates. It would be desirable to obtain predominantly IQC precipitates. In this respect, ageing at 400 °C represents the highest viable temperature.

A lot of precipitates formed during ageing at investigated temperatures, however, the hardness remained more or less the same level as in the as-cast state. This can be contributed to the competition between the solid solution and precipitation strengthening. Nevertheless, typical aluminium alloys lose most of their strengths when heated to temperatures above 250 °C [41], but the room temperature hardness of these alloys remained at high levels after ageing for 240 h at 300 °C and 24 h at 400 °C. In order to reveal the real potential of these alloys for high temperature applications, their high temperature stability should be compared with conventional alloys at the same testing conditions.

5. Conclusions

The main conclusions of this work are:

In the investigated Al-Mn-Cu-Be alloys, the highest number density of spherical IQC precipitates, with the 15–20 nm diameters, can be obtained at 300 °C by T5 treatment.

Increasing ageing temperatures caused formation of coarser IQC precipitates and plate-like decagonal quasicrystalline precipitates.

Higher hardness was attained in the alloy possessing a higher Mn content.

The room temperature hardness of aged alloys remained more or less at the same level as in the as-cast condition.

The precipitates caused strong work hardening during initial plastic deformation up to 0.1 true strain, followed by strain softening at higher strains.

The investigated alloys showed interesting behaviour; however, further investigations are required to reveal the real potential of these experimental alloys, especially for high temperature applications.

6. Patents

The results of this investigation were very important by application of the following patent. ZUPANIČ, Franc, BONČINA, Tonica. Manufacturing of high strength and heat resistant aluminium alloys strengthened by dual precipitates = Herstellung von Hochfesten und wärmebeständigen durch dual-präzipitate verstärkten Aluminiumlegierungen = Fabrication d'alliages d'aluminium à haute résistance mécanique et thermique renforcés par des précipités doubles: European patent specification EP 3 456 853 B1, 2020-02-19. Munich: European Patent Office, 2020.

Author Contributions: Conceptualization, T.B. and F.Z.; methodology, T.B.; formal analysis, T.B.; investigation, T.B. and M.A.; writing—original draft preparation, T.B.; writing—review and editing, T.B. and M.A.; visualization, T.B. and M.A.; supervision, F.Z. All authors have read and agreed to the published version of the manuscript.

Funding: The XRD investigations were carried out at Elettra, Sincrotrone Trieste, Italy, in the framework of Proposals 20160145 and 20165196. The TEM investigations at TU Graz were enabled by the ESTEEM 3 Project, which has received funding from the European Union's HORIZON 2020 RESEARCH AND INNOVATION PROGRAMME UNDER GRANT AGREEMENT NO 823717. The authors acknowledge the financial support from the Slovenian Research Agency (research core funding No. P2-0120 and P2-0123).

Acknowledgments: We acknowledge Luisa Barba for support by XRD experiments at Elettra, Christian Gspan for S/TEM investigations, and Jaka Burja for compression tests.

Conflicts of Interest: The authors declare no conflict of interest.

References

1. Shechtman, D.; Blech, I.; Gratias, D.; Cahn, J.W. Metallic phase with long-range orientational order and no translational symmetry. *Phys. Rev. Lett.* **1984**, *53*, 1951–1953. [CrossRef]

2.   Schurack, F.; Eckert, J.; Schultz, L. Synthesis and mechanical properties of cast quasicrystal-reinforced Al-alloys. *Acta Mater.* **2001**, *49*, 1351–1361. [CrossRef]

3.   Huo, S.; Mais, B. Characteristics of heat resistant nanoquasicrystalline PM aluminum materials. *Met. Powder Rep.* **2017**, *72*, 45–50. [CrossRef]

4.   Inoue, A.; Kimura, H. Fabrications and mechanical properties of bulk amorphous, nanocrystalline, nanoquasicrystalline alloys in aluminum-based system. *J. Light Met.* **2001**, *1*, 31–41. [CrossRef]

5.   Kim, K.B.; Xu, W.; Tomut, M.; Stoica, M.; Calin, M.; Yi, S.; Lee, W.H.; Eckert, J. Formation of icosahedral phase in an Al93Fe3Cr2Ti2 bulk alloy. *J. Alloys Compd.* **2007**, *436*, L1–L4. [CrossRef]

6.   Kim, S.H.; Song, G.S.; Fleury, E.; Chattopadhyay, K.; Kim, W.T.; Kim, D.H. Icosahedral quasicrystalline and hexagonal approximant phases in the Al-Mn-Be alloy system. *Philos. Mag. A Phys. Condens. Matter Struct. Defects Mech. Prop.* **2002**, *82*, 1495–1508. [CrossRef]

7.   Schurack, F.; Eckert, J.; Schultz, L. Al-Mn-Ce quasicrystalline composites: Phase formation and mechanical properties. *Philos. Mag.* **2003**, *83*, 807–825. [CrossRef]

8.   Boncina, T.; Markoli, B.; Zupanic, F. Characterization of cast Al86Mn3Be11 alloy. *J. Microsc.-Oxf.* **2009**, *233*, 364–371. [CrossRef] [PubMed]

9.   Rouxel, B.; Ramajayam, M.; Langan, T.J.; Lamb, J.; Sanders, P.G.; Dorin, T. Effect of dislocations, Al3(Sc,Zr) distribution and ageing temperature on θ′ precipitation in Al-Cu-(Sc)-(Zr) alloys. *Materialia* **2020**, *9*, 100610. [CrossRef]

10.   Babaniaris, S.; Ramajayam, M.; Jiang, L.; Langan, T.; Dorin, T. Tailored precipitation route for the effective utilisation of Sc and Zr in an Al-Mg-Si alloy. *Materialia* **2020**, *10*, 100656. [CrossRef]

11.   Vončina, M.; Medved, J.; Kores, S.; Xie, P.; Schumacher, P.; Li, J. Precipitation microstructure in Al-Si-Mg-Mn alloy with Zr additions. *Mater. Charact.* **2019**, *155*, 109820. [CrossRef]

12.   Mikhaylovskaya, A.V.; Mochugovskiy, A.G.; Levchenko, V.S.; Tabachkova, N.Y.; Mufalo, W.; Portnoy, V.K. Precipitation behavior of L12 Al3Zr phase in Al-Mg-Zr alloy. *Mater. Charact.* **2018**, *139*, 30–37. [CrossRef]

13.   Hansen, V.; Gjønnes, J. Quasicrystals in an aluminium alloy matrix and the transformation to α-AlMnSi via intermediate stages. *Philos. Mag. A* **1996**, *73*, 1147–1158. [CrossRef]

14.   Li, Y.J.; Arnberg, L. Quantitative study on the precipitation behavior of dispersoids in DC-cast AA3003 alloy during heating and homogenization. *Acta Mater.* **2003**, *51*, 3415–3428. [CrossRef]

15.   Mochugovskiy, A.; Tabachkova, N.; Mikhaylovskaya, A. Annealing induced precipitation of nanoscale icosahedral quasicrystals in aluminum based alloy. *Mater. Lett.* **2019**, *247*, 200–203. [CrossRef]

16.   Boncina, T.; Zupanic, F. In situ tem study of precipitation in a quasicrystal-strengthened al-alloy. *Arch. Metall. Mater.* **2017**, *62*, 5–9. [CrossRef]

17.   Zupanič, F.; Wang, D.; Gspan, C.; Bončina, T. Precipitates in a quasicrystal-strengthened Al–Mn–Be–Cu alloy. *Mater. Charact.* **2015**, *106*, 93–99. [CrossRef]

18.   Shen, Z.; Liu, C.; Ding, Q.; Wang, S.; Wei, X.; Chen, L.; Li, J.; Zhang, Z. The structure determination of Al20Cu2Mn3 by near atomic resolution chemical mapping. *J. Alloys Compd.* **2014**, *601*, 25–30. [CrossRef]

19.   Zupanič, F.; Gspan, C.; Burja, J.; Bončina, T. Quasicrystalline and L12 precipitates in a microalloyed Al-Mn-Cu alloy. *Mater. Today Commun.* **2020**, *22*, 100809. [CrossRef]

20.   Bončina, T.; Zupanič, F. Methods of Deep Etching and Particle Extracting for Controlling of Aluminium Alloys. *Mater. Today Proc.* **2019**, *10*, 248–254. [CrossRef]

21.   Predel, B. Al-Mn (Aluminum-Manganese). In *Ac-Ag....Au-Zr: Supplement to Subvolume IV/5A (Landolt-Börnstein: Numerical Data and Functional Relationships in Science and Technology—New Series) (No. 4)*; Springer: Berlin/Heidelberg, Germany, 2006; ISBN 978-3540435341.

22.   Predel, B. Al-Cu (Aluminum-Copper). In *Ac-Ag....Au-Zr: Supplement to Subvolume IV/5A (Landolt-Börnstein: Numerical Data and Functional Relationships in Science and Technology—New Series) (No. 4)*; Springer: Berlin/Heidelberg, Germany, 2006; ISBN 978-3540435341.

23.   Stan-Glowinska, K. Formation of Quasicrystalline Phases and Their Close Approximants in Cast Al-Mn Base Alloys Modified by Transition Metals. *Crystals* **2018**, *8*. [CrossRef]

24.   Rozman, N.; Medved, J.; Zupanic, F. Microstructural evolution in Al-Mn-Cu-(Be) alloys. *Philos. Mag.* **2011**, *91*, 4230–4246. [CrossRef]

25.   Elser, V. Indexing problems in quasicrystal diffraction. *Phys. Rev. B* **1985**, *32*, 4892–4898. [CrossRef] [PubMed]

26.   Singh, A.; Ranganathan, S. On the indexing and reciprocal space of icosahedral quasicrystal. *J. Mater. Res.* **1999**, *14*, 4182–4187. [CrossRef]

27. Fung, K.K.; Zhou, Y.Q. Direct observation of the transformation of the icosahedral phase in (Al6Mn)1−xSix, into α(AlMnSi). *Philos. Mag. B* **1986**, *54*, L27–L31. [CrossRef]

28. Boncina, T. Shapes of the icosahedral quasicrystalline phase in melt-spun ribbons. *Metalurgija* **2013**, *52*, 65–67.

29. Chattopadhyay, K.; Ravishankar, N.; Goswami, R. Shapes of quasicrystals. *Prog. Cryst. Growth Charact. Mater.* **1997**, *34*, 237–249. [CrossRef]

30. Andersen, S.J.; Marioara, C.D.; Friis, J.; Wenner, S.; Holmestad, R. Precipitates in aluminium alloys. *Adv. Phys. X* **2018**, *3*, 1479984. [CrossRef]

31. Zupanic, F.; Markoli, B.; Naglic, I.; Weingartner, T.; Meden, A.; Boncina, T. Phases in the Al-Corner of the Al-Mn-Be System. *Microsc. Microanal.* **2013**, *19*, 1308–1316. [CrossRef]

32. Lukas, H.L. Al-Cu-Mn (Aluminium-Copper-Manganese). In *Light Metal Systems. Part 2*; Effenberg, G., Ilyenko, S., Eds.; Springer: Berlin/Heidelberg, Germany, 2005; Volume 11A2. Available online: https://materials.springer.com/lb/docs/sm_lbs_978-3-540-31687-9_5 (accessed on 10 July 2020). [CrossRef]

33. Belov, N.A.; Alabin, A.N.; Matveeva, I.A. Optimization of phase composition of Al-Cu-Mn-Zr-Sc alloys for rolled products without requirement for solution treatment and quenching. *J. Alloys Compd.* **2014**, *583*, 206–213. [CrossRef]

34. Singh, A.; Somekawa, H.; Tsai, A.P. Interfaces made by tin with icosahedral phase matrix. *Scr. Mater.* **2008**, *59*, 699–702. [CrossRef]

35. Singh, A.; Somekawa, H.; Matsushita, Y.; Tsai, A.P. Solidification of tin on quasicrystalline surfaces. *Philos. Mag.* **2012**, *92*, 1106–1128. [CrossRef]

36. Singh, A.; Tsai, A.P.; Nakamura, M.; Watanabe, M.; Kato, A. Nanoprecipitates of icosahedral phase in quasicrystal-strengthened Mg-Zn-Y alloys. *Philos. Mag. Lett.* **2003**, *83*, 543–551. [CrossRef]

37. Mikhaylovskaya, A.V.; Kishchik, A.A.; Kotov, A.D.; Rofman, O.V.; Tabachkova, N.Y. Precipitation behavior and high strain rate superplasticity in a novel fine-grained aluminum based alloy. *Mater. Sci. Eng. A* **2019**, *760*, 37–46. [CrossRef]

38. Du, Y.; Chang, Y.A.; Huang, B.Y.; Gong, W.P.; Jin, Z.P.; Xu, H.H.; Yuan, Z.H.; Liu, Y.; He, Y.H.; Xie, F.Y. Diffusion coefficients of some solutes in fcc and liquid Al: Critical evaluation and correlation. *Mater. Sci. Eng. A Struct. Mater. Prop. Microstruct. Process.* **2003**, *363*, 140–151. [CrossRef]

39. Christian, J.W. CHAPTER 16—Precipitation from Supersaturated Solid Solution. In *The Theory of Transformations in Metals and Alloys*; Christian, J.W., Ed.; Pergamon: Oxford, UK, 2002; pp. 718–796.

40. Štrekelj, N.; Naglič, I.; Klančnik, G.; Nagode, A.; Markoli, B. Microstructural changes in quasicrystalline Al–Mn–Be–Cu alloy after various heat treatments. *Int. J. Mater. Res.* **2015**, *106*, 342–351. [CrossRef]

41. Kaufman, J.G. *Properties of Aluminium Alloys—Tensile, Creep and Fatigue Data at High and Low Temperatures*; ASM International: Metals Park, OH, USA, 1999.

 *metals* 

Article

Effects of Homogenization Conditions on the Microstructure Evolution of Aluminium Alloy EN AW 8006

Maja Vončina [1,*], Kristijan Kresnik [1], Darja Volšak [2] and Jožef Medved [1]

[1] University of Ljubljana, Faculty of Natural Sciences and Engineering, Department for Materials and Metallurgy, Aškerčeva 12, Ljubljana, Slovenia; kristijankresnik@gmail.com (K.K.); jozef.medved@omm.ntf.uni-lj.si (J.M.)

[2] Impol Aluminium Industry d.d., Partizanska 38, 2310 Slovenska Bistrica, Slovenia; darja.volsak@impol.si

* Correspondence: maja.voncina@omm.ntf.uni-lj.si; Tel.: +386-1-200-0418

Received: 28 February 2020; Accepted: 22 March 2020; Published: 24 March 2020

Abstract: The industrial production of products, such as foil and aluminium alloy strips, begins with the production of semi-finished products in the form of slabs. These are produced by the continuous casting process, which is quick and does not allow the equilibrium conditions of solidification. Non-homogeneity—such as micro and macro segregation, non-equilibrium phases and microstructural constituents, as well as stresses arising during non-equilibrium solidification—are eliminated by means of homogenization annealing. In this way, a number of technological difficulties in the further processing of semi-finished products can be avoided. The aim of this research was the optimization of the homogenization annealing of the EN AW 8006 alloy. With the Thermo-Calc software, a thermodynamic simulation of equilibrium and non-equilibrium solidification was performed. Differential scanning calorimetry (DSC) was performed on selected samples in as-cast state and after various regimes of homogenization annealing and was used for the simulation of homogenization annealing. Using an optical microscope (OM), a scanning electron microscope (SEM) and an energy dispersion spectrometer (EDS), the microstructure of the samples was examined. Based on the results, it was concluded that homogenization annealing has already taken place after 8 h at 580 °C to the extent, that the material is then suitable for further processing.

Keywords: wrought aluminium alloy; homogenization annealing; thermodynamic equilibrium; microstructure; intermetallic phases; differential scanning calorimetry (DSC)

---

1. Introduction

Due to their relatively high strength, combined with their sufficient ductility, the non-hardenable alloys Al-Fe-Mn-Si are largely used in the automotive industry, cooling systems, civil engineering, etc. Al-foil can be produced either by starting from a conventional direct chill (DC) cast ingot which is then hot rolled to a strip of about 2–5 mm, or from continuous casting a 6–7-mm-thick sheet (twin roll casting or belt casting) and then cold rolling it to an intermediate (foil stock) gauge of about 0.4–1 mm. This is followed by annealing in the temperature range 350–400 °C. [1] Then, the material is cold rolled to final foil thickness. Afterwards, closed gap rolling must be used to achieve thinner foil gauges below 100 µm, which is performed in a dedicated foil mill. Most Al-foil products will be used in the O-temper conditions (foil must be soft annealed). In addition to the recrystallization, the final gauge annealing is required to remove lubricant from the foil [2].

Thin aluminium sheets which exhibit good formability and sufficient rigidity are advisable for use in applications such as fin-stocks in heat exchangers. Thus, a fine grain size structure and a texture warranting low anisotropy must be achieved by alloy processing. The form in which alloying elements are present is an important factor influencing recrystallization and texture development, and in consequence alloy properties. Iron, manganese and silicon may be present in second-phase particles or dissolved in the matrix. Coarse intermetallic particles usually act as nucleation sites during recrystallization, giving a random or retained rolling texture. Fine particles inhibit grain growth by pinning grain boundaries. Thus, depending on dispersoid density and size, recrystallization can be accelerated or retarded. Alloying elements in a solid solution also affects structure and properties. Solid solution element content and dispersoid distribution can be changed by heat treatment [3–5].

Detailed knowledge of the correlation of the microstructure and processing parameters is required for proper control of the final material's properties. [6,7]. In relatively clean Al-foil alloys, especially the material's microchemistry, i.e., the solution/precipitation state of the alloying elements Fe and Si, and other impurities, will affect both processing and foil properties at the final gauge. Commercial Al-alloys will always comprise large Fe-bearing phases because of the very low solubility of Fe in Al. The type, volume, size, and especially the morphology of these constituent particles have an impact on ductility and formability. Pre-heating or homogenization annealing prior to hot rolling will affect both processing and final foil properties by leading to the formation of fine secondary intermetallic phases or so-called 'dispersoids' [6,7]. The solute level upon hot deformation must be kept low, whereas there is a limitation to the amount of strengthening through solid solution hardening, such that the dominant strengthening mechanism in dilute Al-foil alloys is dispersion hardening, with some contribution from solutes. Hence, the volume fraction and size distribution of the dispersoids is an important factor in dispersion strengthening [8,9]. Increased hardness leads to reduced formability and fatigue resistance as a result of needle-shaped particles. Grain size has a relatively small influence on strength, but largely impacts ductility, with finer grain size giving higher ductility [6,10].

Figure 1a shows the Al-rich corner of the equilibrium ternary phase diagram Al-Fe-Si as a function of the Si and Fe content for the investigated alloy EN AW 8011. [11,12] The diagram displays a total of three different second-phase particles in equilibrium with the Al-matrix (α-Al), for which the stability ranges depend on temperature and the exact Si content. The maximum solubility of Fe in Al is 0.05 wt. % at 650 °C. Accordingly, alloys based on commercial purity will always comprise Fe-bearing constituent phases. Alloys with low Si content usually contain rather large fractions of the phase Al_3Fe (or $Al_{13}Fe_4$). This phase adversely affects the formability of Al-foil alloys by forming needle-like particles [13]. Al-alloys with increased Si content over 0.5 wt. % display significant portions of ternary AlFeSi-phases [12]. At medium Si contents and/or high temperatures, the α-AlFeSi phase is obtained, whereas the chemical composition of this phase shows rather large scatter and is described as Al_8Fe_2Si or $Al_{12}Fe_3Si$. The ratio of Fe:Si may vary between 2:1 and 3:1 in at. % or, between 4:1 and 6:1 in wt. %. Increased Si contents and lower temperatures tend to stabilize the phase β-AlFeSi. For this phase a stoichiometric composition of Al_5FeSi is reported; thus, the Fe:Si ratio in at. % is 1:1, and in wt. % this ratio is very close to 2:1 [1].

Analysis of EN AW 8006 alloys at a low content of Si impurity can be considered using an Al-Fe-Mn phase diagram (Figure 1b). The isothermal sections of this diagram in the solid state have only one three phase region (Al) + Al_6(FeMn) + Al_3Fe. Polythermal sections within the compositional range of EN AW 8006 commercial alloys have only one invariant horizontal. The sections at 0.7 wt. % Mn and 1.6 wt. % Fe show that primary crystals of the Al_3Fe and Al_6(FeMn) phases can form when the Fe concentration is at the upper limit [14].

Due to dispersion hardening and grain boundary hardening, strain hardening capacity increases simultaneously with an increasing Fe content, which leads to an increase of the tensile strength and a decrease in the grain size, and therefore produces high elongation values. The castability improves from the addition of Si and, furthermore, the cast structure becomes more uniform. Si accelerates the precipitation of dissolved solute elements during annealing. Si contents higher than 0.8 wt. % lower the recrystallization temperature, limit the final anneal temperature range, and reduce the positive effect of Fe on the grain size. Generally, Mn retards recrystallization and increases the recrystallization temperature [15].

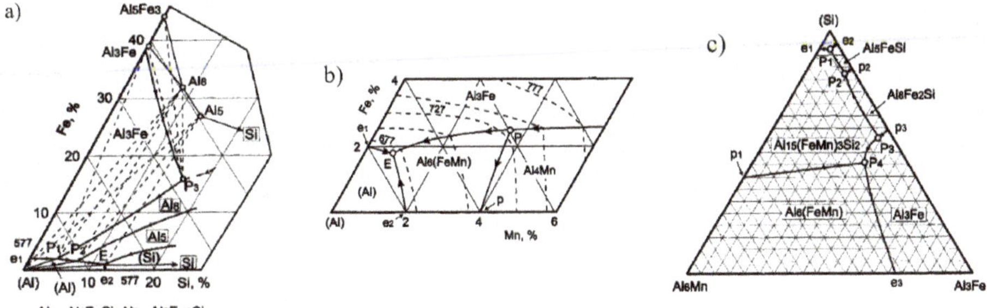

Figure 1. Phase diagram of an Al-Fe-Si system (**a**), a liquidus projection of an Al-Fe-Mn system (**b**), and a polythermal projection of the solidification surface of an Al-Fe-Mn-Si system (**c**) reproduces from reference [12].

The conditions for equilibrium solidification are not fulfilled in practice due to quite high cooling rates prevailing during industrial DC-casting [9]. Therefore, the as-cast ingot will be supersaturated with Fe and by other slow-diffusing species, such as Mn and Cr, if they are present. Under these conditions, a metastable phase Al$_6$Fe can form in addition to the stable monoclinic Al$_3$Fe phase. This phase is isomorphous to the orthorhombic Al$_6$Mn phase usually encountered in Mn-containing alloys (3xxx or 5xxx series alloys) [12]. Furthermore, the literature describes other metastable phases of Al$_m$Fe where m varies from 4 to 5 [14]. A transformation of the metastable Fe-bearing phases towards stable Al$_3$Fe can be achieved by homogenization at high temperatures [7,16,17]. If Mn is present, the α-AlFeSi phase is replaced by an isostructural quaternary α-Al(Mn,Fe)Si phase from the system presented in Figure 1c, which may show large variation in exact chemical composition [6,16].

The industrial production of products, such as strips and foils, from aluminium alloys starts with a semi-finished product in the form of a slab, which does not allow solidification at equilibrium conditions. Inhomogeneities, micro-segregations, residual stresses and other defects that occur in the material during un-equilibrium solidification are eliminated with homogenization annealing. [1] The more present internal defects are the large AlFeMn and/or AlFeMnSi intermetallic particles with different sizes, which are aligned on the deformation direction. Some zones have a coarse and fragile compound. The agglomeration of intermetallic particles leads to the rupture of the foil. Heat treatment (the homogenization of the alloy at 570 °C/180 min.) can prevent the structure defects observed in sheets and foils produced from EN AW 8006 alloy. [18,19] Aluminium oxide and/or spinel (MgO·Al$_2$O$_3$) on the surface of the sheets can be observed. It is more likely for these inclusions to become sources of pinholes in the sheets. Moreover, spherical particles of aluminium boride can be seen on the surface of the sheets. They are small in size and non-deformable particles compared to the ductile aluminium matrix, but they can affect the performance of the sheet rolling [19,20].

Homogenization processing has a critical contribution by allowing the precipitation of excessive alloying elements in the solid solution before subsequent thermomechanical processing. However, the high solidification rate encountered in TRC (twin-roll casting) not only favours the supersaturation and metastable condition of the alloying elements, but also hinders their micro-segregation [21,22].

In the present study, the microstructural evolution of the typical dilute Al-Fe-Si (EN AW 8006) foil alloy during homogenization was investigated using optical and electron microscopy as well as differential scanning calorimetry (DSC). The effect of different homogenization practices on the formation of intermetallic phases was analysed. Especially, the impact of homogenization on changes in the morphology of plate-like constituent particles which are known [23] to adversely affect the formability of 8xxx series alloys was examined.

2. Materials and Methods

The aim of the study was to optimize the homogenization annealing of the EN AW 8006 alloy slab. Standard composition is listed in Table 1.

Table 1. The standard chemical composition of the EN AW 8006 alloy/wt. %.

Element	Si	Fe	Cu	Mn	Mg	Zn	Other	Al
Amount	0.40	1.2–2.0	0.3	0.30–1.0	0.10	0.10	0.05	Bal.

A thermodynamic analysis was conducted using Thermo-Calc Software (TCW2019, Stockholm, Sweden) and TCAL6 database to obtain the phase diagram and the stability of different phases, as well as the equilibrium concentrations of these phases.

Samples for heat treatment investigations were cut from the middle of the front surface of a slab with dimensions of 510 × 1310 × 4800 mm, in the form of cubes. The samples' dimensions were approximately 15 mm (l) × 15 mm (w) × 15 mm (h). Single-step solution heat treatments (homogenization treatments) were done using an electric chamber furnace, which was previously calibrated. For the investigation of solution heat treatments, samples were held for various times—4 h, 6 h, 8 h, 10 h and 12 h—at temperatures of 580 °C and 600 °C to investigate the effect of homogenization on the dissolution of particles. Samples were, after a certain time, removed from the furnace and cooled in air in order to simulate industrial conditions. All metallographic examinations were carried out on half of the samples after homogenization annealing, whereas the remainder were grinded and polished according to the standard metallographic procedure for aluminium alloys. Differential scanning calorimetry (DSC) was performed on samples in an as-cast state and on the second half of the homogenized samples, after different regimes of homogenization annealing, using an STA Jupiter 449c apparatus (NETZSCH Group, Selb, Germany). DSC analysis was performed as follows: heating up to 710 °C, followed by cooling to room temperature using 10 K/min heating and cooling rate in a protective atmosphere of argon. Using an optic microscope Zeiss Axio Observer 7 (ZEISS International, Jena, Germany) and a scanning electron microscope SEM—Jeol JSM 6610LV (JEOL Ltd., Tokyo, Japan) with an energy dispersion spectrometer (EDS), the microstructural components were analysed.

Samples for homogenisation simulation on the DSC apparatus were taken from the middle of the slab. On the DSC device, the homogenization of the as-cast samples was carried out, which lasted for 12 h at 580 °C and 600 °C. By using the tangent method, the change in the slope of the DSC curves was determined, which is also attributed to changes in the course of homogenization.

3. Results and discussion

Figure 2 illustrates the results of the thermodynamic calculations. The results indicate that the temperature range for the dissolution is 550–620 °C. As shown in Figure 2, the equilibrium phases and the temperature ranges in which they are present are: the $Al_{13}Fe_4$ phase up to 652 °C, the $Al_{15}Si_2Mn_4$ phase up to 630 °C, the $Al_9Fe_2Si_2$ phase up to 477 °C, and the Al_6Mn phase up to 444 °C. This implies that by heating at 580 °C or 600 °C for a long holding time, $Al_{15}Si_2Mn_4$ will partially dissolve, and $Al_9Fe_2Si_2$ and Al_6Mn will completely dissolve, but the iron phase $Al_{13}Fe_4$, as well as a fraction of the $Al_{15}Si_2Mn_4$ phase, will still be present. It should also be noted that the homogenization treatment for these alloys should not be conducted at temperatures over 640 °C, as the Fe-rich eutectic phase $Al_{13}Fe_4$ will start to melt.

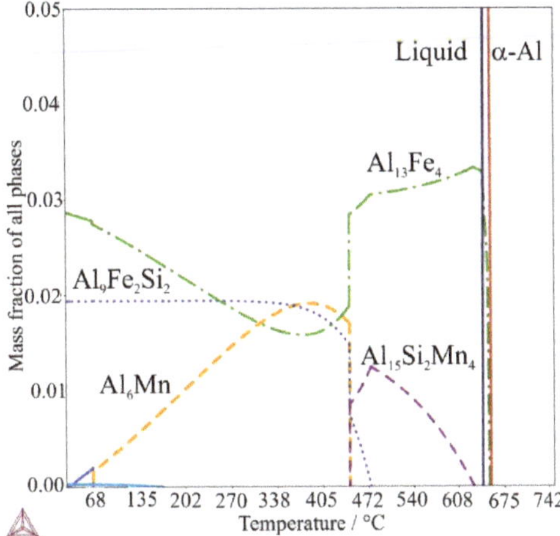

Figure 2. Thermo-Calc predicted phase stability in aluminium alloy EN AW 8006 with regard to the temperature.

In Figure 3, the heating and cooling DSC curves of the experimental samples are presented, whereas only samples in an as-cast state, and after 6 h and 12 h at temperatures of 580 °C and 600 °C, were analysed.

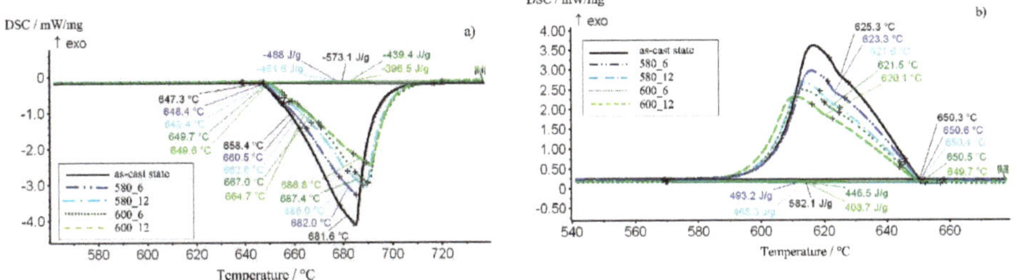

Figure 3. Heating (**a**) and cooling (**b**) Differential scanning calorimetry (DSC) curves of the investigated samples after homogenization.

From these results, characteristic temperatures at heating and cooling, as well as melting (ΔH_m) and solidification (ΔH_s) enthalpies were determined.

From Figure 3a it can be determined, according to Thermo-Calc calculations, that at a temperature of about 649 °C the melting of the eutectic with the Mn-phase occurs, at a temperature of about 664 °C the melting of eutectic (α-Al + Al$_{13}$Fe$_4$) occurs, and at a temperature of 685 °C the melting of the primary crystals α-Al takes place. From the cooling DSC curves, the solidification is as follows (Figure 3b): the primary solidification of α-Al occurs at about 650 °C, and the solidification of the eutectic (α-Al + Al$_{13}$Fe$_4$) occurs at a temperature of about 621 °C. The solidification of Mn-phase could not be detected due to an equilibrium solidification.

In order to improve transparency, the temperatures and enthalpies are collected in Table 2. It can be considered that the melting temperature of the Mn-eutectic is the lowest in the sample in the as-cast state (647.3 °C), the melting temperature is increased in homogenized samples, and is the highest (649.7 °C) in samples that were homogenized at 600 °C. This trend is also observed in the second eutectic (α-Al + Al$_{13}$Fe$_4$). The melting temperature of the α-Al phase is also the lowest in the sample in the as-cast state (681.6 °C), which results from a slower and more equilibrious solidification. It is raised by a higher homogenization temperature and a longer homogenization annealing time. It is also observed that the largest melting enthalpy is in the sample in the as-cast state (-573.1 J/g). Homogenized samples have a lower melting enthalpy, where the lowest is in the sample after homogenization at 600 °C 12 h.

Table 2. Listed melting and solidification temperatures, as well as corresponding enthalpies.

Sample	Heating/Melting				Cooling/Solidification		
	ΔH_m (J/g)	$T_{Al15Si2Mn4}$ (°C)	$T_{Al13Fe4}$ (°C)	$T_{\alpha\text{-Al}}$ (°C)	ΔH_s (J/g)	$T_{\alpha\text{-Al}}$ (°C)	$T_{Al13Fe4}$ (°C)
As cast	-73.1	647.3	658.4	681.6	582.1	650.3	625.3
580-6	-488	648.4	660.5	682.0	493.2	650.6	623.3
580-12	-461.6	649.4	662. 8	685.0	486.3	650.4	621.6
600-6	-439.4	649.7	667.0	687.4	446.5	650.5	621.5
600-12	-396.5	649.6	664.7	686.8	408.7	649.7	620.1

The liquidus temperatures of all samples are around 650 °C. At cooling, only the primary solidification and the solidification of the eutectic (α-Al + $Al_{13}Fe_4$) can be observed on the cooling DSC curve, which results from a slower and more equilibrious solidification. The liquidus temperature does not change significantly with the time and temperature of the prior homogenization annealing. The highest solidification temperature of the eutectic (α-Al + $Al_{13}Fe_4$) is in the sample in the cast state (625.3 °C), which decreases with increasing homogenization temperature and time. The tendency of the solidification enthalpy variation is the same as in the melting enthalpy.

These observations correlate with the fact that the effects of homogenization are greater for longer periods and higher annealing temperatures.

The optical micrographs taken from samples after various homogenization regimes are shown in Figures 4 and 5. In the as-cast sample, there is at most a finely dispersed eutectic that occurs during longer $Al_{13}Fe_4$ needles. A slightly larger square phase of $Al_{15}Si_2Mn_4$ irregular forms is also observed. In homogenized samples, dispersed phases, which look like $Al_{13}Fe_4$, appear in the α-Al matrix. The eutectic phase $Al_{15}Si_2Mn_4$ gradually dissolves and disappears with longer annealing times. Those which remain become slightly larger and rounded, which is desirable due to their lower cut effect and thus their better forming properties. The longer needle phases $Al_{13}Fe_4$ become more rounded by increasing the annealing time. Similarly, it occurs at both temperatures, but at a higher temperature of annealing (Figure 5) the less eutectic phase is seen, which is somewhat more rounded. The content and shape of the intermetallic phases does not change significantly after eight hours of annealing.

In addition, the simulation of the homogenization was done using DSC measurement, whereas the homogenization process was analysed at two experimental temperatures, 590 °C and 610 °C, respectively. Each experiment was performed two times; the results are presented in Figure 6. The results show that when homogenization takes place at 590 °C, not all phases are dissolved in 12 h, but rather that the process is continuing, whereas the DSC curve is still dropping. At 610 °C, most of the homogenization process is finished after 160 min, and fully finished after 300 min, at 610 °C. These results are in good agreement with the microstructure results in Figures 4 and 5.

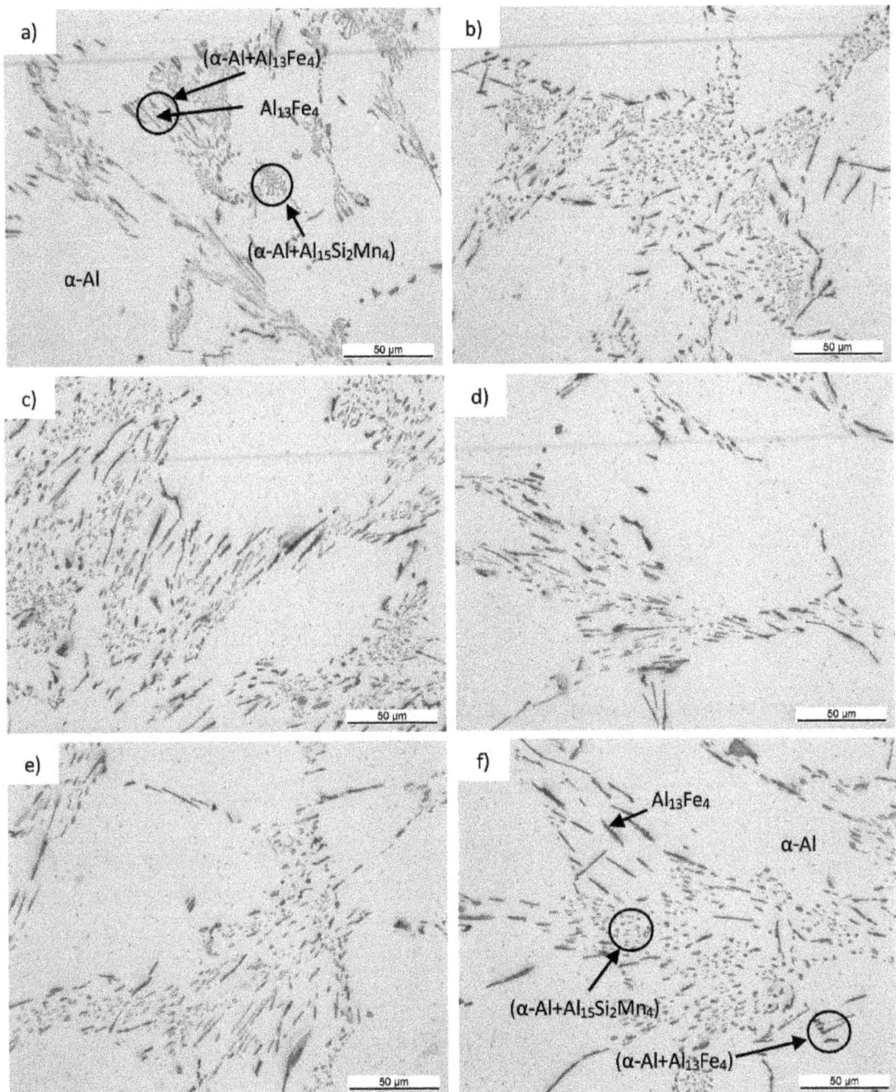

Figure 4. Microstructure of a sample in an as-cast state, and of homogenized samples at 580 °C, at 500× magnification: in an as-cast state (**a**) and after 4 h (**b**), 6 h (**c**), 8 h (**d**), 10 h (**e**) and 12 h (**f**).

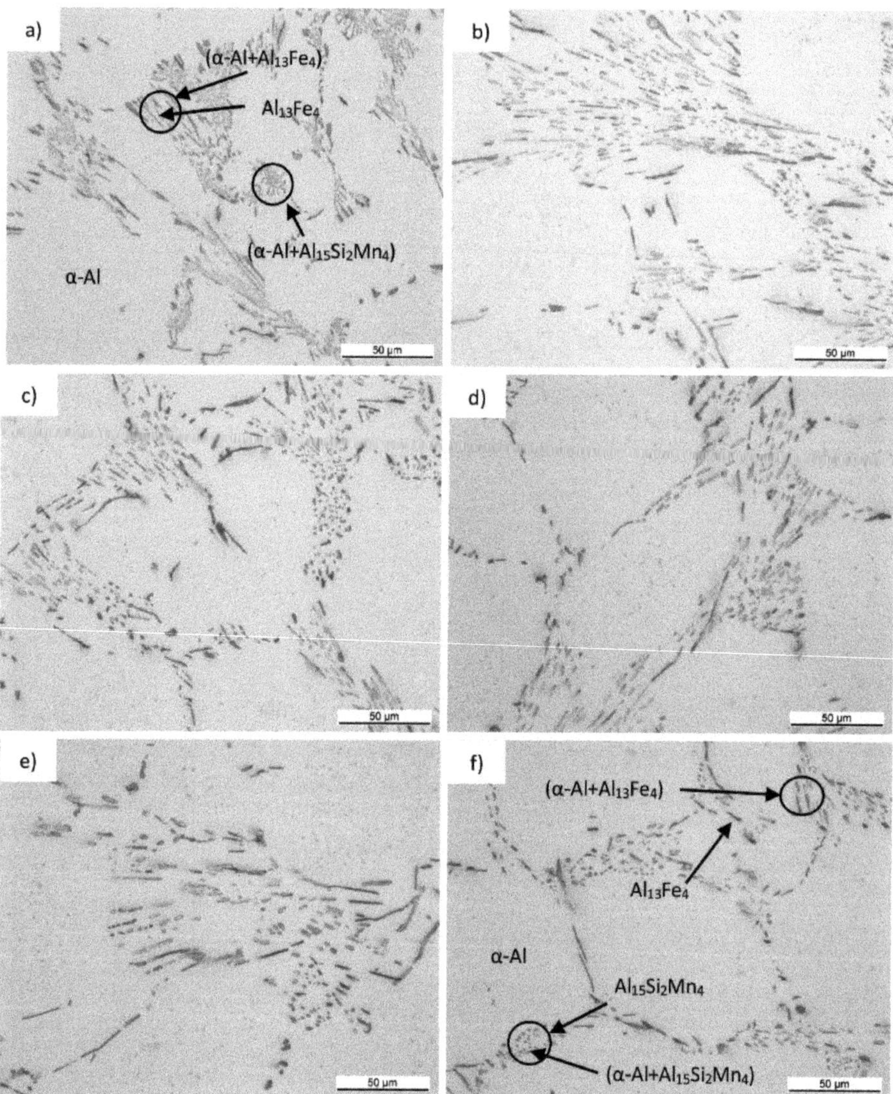

Figure 5. Microstructure of sample in as-cast state and homogenized samples at 600 °C at 500× magnification: in as-cast state (**a**) and after 4 h (**b**), 6 h (**c**), 8 h (**d**), 10 h (**e**) and 12 h (**f**).

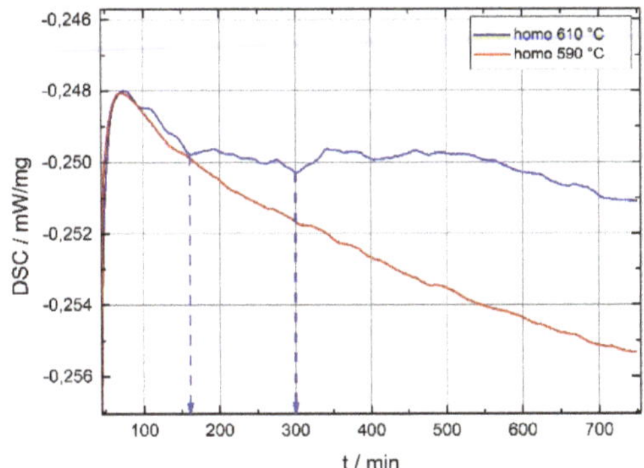

Figure 6. DSC measurement of homogenization annealing of EN AW 8006 alloy at 590 °C and 610 °C.

Furthermore, an EDS analysis of the selected samples (Figure 7) was made in order to determine the type and phase composition of phases after certain homogenization regimes. From the EDS analysis, listed in Table 3, it can be concluded that the phases in the form of large, sharp needles contain mainly aluminium and iron, which indicates that this is the $Al_{13}Fe_4$ phase. Phases of smaller and rounded forms contain a slightly higher concentration of manganese and silicon, indicating that this is the $Al_{15}Si_2Mn_4$ phase. The proportion of these two elements increases in these phases with a longer period of homogenization annealing and is in the as-cast sample up to 2 wt. %, and in the sample 580–12 as well as up to 4 mas. %.

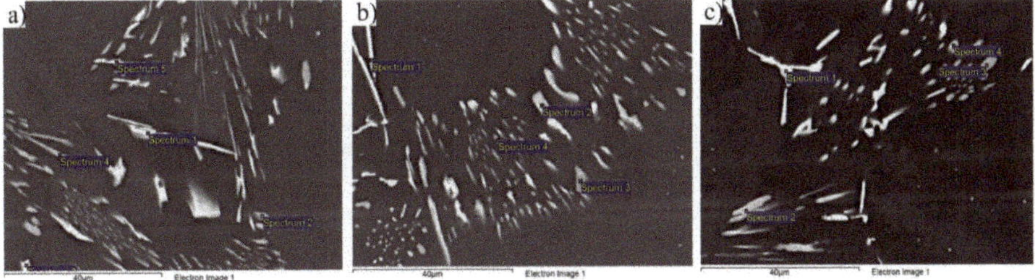

Figure 7. SEM (scanning electron microscope) micrographs of the as-cast sample (**a**), and the samples homogenized for 6 h (**b**) and 12 h (**c**) at a temperature of 580 °C, wherein EDS (energy dispersion spectrometer) analysed spots are marked.

Table 3. EDS results showing different phases present in the investigated samples from Figure 7 in at. %.

State	As-Cast					580-6				580-12			
Element	1	2	3	4	5	1	2	3	4	1	2	3	4
Al	77.96	87.36	78.47	91.1	83.03	82.05	78.92	88.58	91.08	79.01	79.68	85.65	82.76
Si	1.43	0.34	1.39	0.22	1.03	0.71	3.93	2.91	0.15	0.81	4.43	1.2	4.15
Mn	1.14	1.5	1.27	1.1	0.82	0.86	2.42	2.16	1.75	1.13	3.07	2.47	2.84
Fe	19.46	10.8	18.87	7.59	15.11	16.38	14.74	6.35	7.02	19.05	12.82	10.69	10.25
Estimated phase	$Al_{15}Si_2Mn_4$	-	$Al_{15}Si_2Mn_4$	$Al_{13}Fe_4$	$Al_{13}Fe_4$	$Al_{13}Fe_4$	$Al_{15}Si_2Mn_4$	$Al_{15}Si_2Mn_4$	-	$Al_{13}Fe_4$	$Al_{15}Si_2Mn_4$	$Al_{15}Si_2Mn_4$	$Al_{15}Si_2Mn_4$

Table 4. EDS results showing different phases present in the investigated samples from Figure 8 in at. %.

State	600-4					600-6					600-12				
Element	1	2	3	4	5	1	2	3	4	5	1	2	3	4	5
Al	79.41	86.63	81.55	80.52	90.11	83.04	80.16	79.45	81.15	91.61	88.2	77.44	84.59	81.39	85.21
Si	0.95	0.07	0.88	0.83	0.14	0.79	0.7	0.78	4.72	0.14	0.72	0.77	4.12	0.76	0.55
Mn	1.73	2.47	1.24	1.25	2.06	1.35	1.05	1.29	3.2	2	1.24	1.32	3.02	1.53	1.21
Fe	17.91	10.83	16.33	17.41	7.69	14.82	18.09	18.48	10.94	6.25	17.83	20.47	8.27	15.78	13.04
Estimated phase	$Al_{13}Fe_4$	-	$Al_{13}Fe_4$	$Al_{13}Fe_4$	-	$Al_{13}Fe_4$	$Al_{13}Fe_4$	$Al_{13}Fe_4$	$Al_{15}Si_2Mn_4$	-	$Al_{13}Fe_4$	$Al_{13}Fe_4$	$Al_{15}Si_2Mn_4$	$Al_{13}Fe_4$	$Al_{13}Fe_4$

For samples that were homogenized at 600 °C (Figure 8), the effects of homogenization are even clearer. The needles of the eutectic phase $Al_{13}Fe_4$ grow and become even more pronounced. Mn-phases are slightly less noticeable with a longer homogenization time. They form more roundish, their particles are smaller and accumulate in larger aggregates, whereas the concentration of manganese and silicon increases (Table 4).

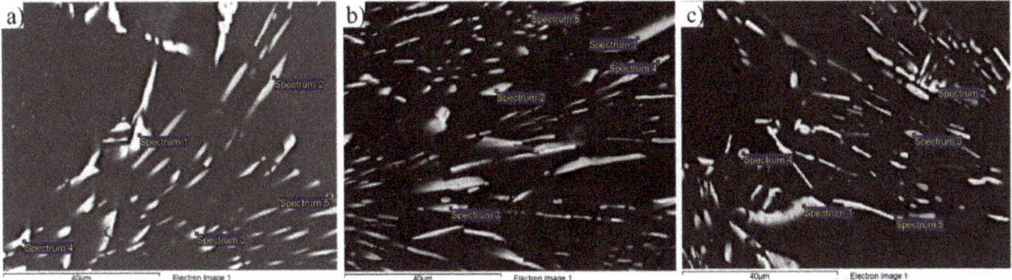

Figure 8. SEM micrographs of samples homogenized for 4 h (**a**), 6 h (**b**) and 12 h (**c**) at a temperature of 600 °C, wherein EDS analysed spots are marked.

4. Conclusions

In this research, the optimization of the homogenization annealing of EN AW 8006 alloy was performed, and the following conclusions were drawn.

In samples of EN AW 8006 alloy, the following phases can occur: α-Al; long needles from the $Al_{13}Fe_4$ phase; finely dispersed (α-Al + $Al_{13}Fe_4$) eutectic; and larger, more square forms of the $Al_{15}Si_2Mn_4$ phase. By prolonging the time of the homogenization annealing, the finely dispersed eutectic (α-Al + $Al_{13}Fe_4$) is partially dissolved in α-Al, and partly forms longer, slightly more rounded needles of the $Al_{13}Fe_4$ phase, which is desirable due to a lower notching effect. The phase $Al_{15}Si_2Mn_4$ also dissolves during homogenization and forms larger globules. It should be noted that the content and shape of the phases after eight hours of annealing does not change significantly at each homogenization temperature.

From the results of EDS phase analysis, it can be concluded that, in the microstructure, there is a solid solution of α-Al and the phase $Al_{13}Fe_4$ and $Al_{15}Si_2Mn_4$, whereas the concentration of silicon and manganese is increased with a longer homogenization time in the $Al_{15}Si_2Mn_4$ phase.

The homogenization annealing time could have been reduced from its current 580 °C for 12 h, wherein preheating takes 5 h, to 6 h at 600 °C, according to DSC and SEM analysis, or to 8 h at 580 °C, according to SEM analysis.

Author Contributions: M.V. carried out the literature review, performed the experiments, analysed the results and wrote the paper. K.K. made the measurements, carried out the metallography and constructed the diagram. D.V. analysed the data and revised the paper. J.M. optimized the research program, improved the idea for the paper, and revised the paper. All authors have read and agreed to the published version of the manuscript.

Funding: This research was supported by Programme P2-0344(B), Functionally graded materials.

Conflicts of Interest: The authors declare no conflict of interest.

References

1. Lentz, M.; Laptyeva, G.; Engler, O. Characterization of second-phase particles in two aluminium foil alloys. *J. Alloy. Compd.* **2016**, *660*, 276–288. [CrossRef]

2. Kucuk, I. Effect of cold rolling reduction rate on corrosion behaviour of twin-roll cast 8006 Aluminium alloys. *Cumhur. Sci. J.* **2018**, *39*, 233–242.

3. Slámová, M.; Očenašek, V.; Dvorak, P.; Juriček, Z. Response of AA 8006 and AA 8111 strip-cast cold rolled alloys to high temperature annealing. *Alum. Alloy.* **1998**, *2*, 1287–1292.

4. Slámová, M.; Očenašek, V.; Dvorak, P.; Juriček, Z. Phase transformation study of two aluminium strip-cast alloys. *Alum. Alloy.* **1998**, *2*, 897–902.

5. Chen, Z.; Zhao, J.; Chen, P. Microstructure and mechanical properties of nanostructured A8006 ribbons. *Mater. Sci. Eng. A* **2012**, *552*, 189–193. [CrossRef]

6. Engler, O.; Laptyeva, G.; Wang, N. Impact of homogenization on microchemistry and recrystallization of the Al-Fe-Mn Alloy AA 8006. *Mater. Charact.* **2013**, *79*, 60–75. [CrossRef]

7. Shakiba, M.; Parson, N.; Chen, X.-G. Effect of homogenization treatment and silicon content on the microstructure and hot work ability of dilute Al–Fe–Si alloys. *Mater. Sci. Eng. A* **2014**, *619*, 180–189. [CrossRef]

8. Roy, R.K.; Kar, S.; Das, S. Evolution of microstructure and mechanical properties during annealing of cold-rolled AA8011 alloy. *J. Alloy. Compd.* **2009**, *468*, 122–129. [CrossRef]

9. Goulart, P.R.; Lazarine, V.B.; Leal, C.V.; Spinelli, J.E.; Cheung, N.; Garcia, A. Investigation of intermetallics in hypoeutectic Al–Fe alloys by dissolution of the Al matrix. *Intermetallics* **2009**, *17*, 753–761. [CrossRef]

10. Slámová, M. Effect of strain level on recrystallisation response of AA8006 and AA8011 thin strips. In Proceedings of the METAL 2001, Ostrava, Czech Republic, 15–17 May 2001.

11. Bale, C.W.; Chartrand, P.; Degterov, S.A.; Eriksson, G.; Hack, K.; Mahfoud, B.R.; Melançon, J.; Pelton, A.D.; Petersen, S. Factsage thermochemical software and databases. *Calphad* **2002**, *26*, 189–228. [CrossRef]

12. Belov, N.A.; Eskin, D.G.; Aksenov, A.A. *Multicomponent Phase diagrams, Applications for Commercial Aluminum Alloys*; Elsevier: Amsterdam, The Netherlands, 2005; p. 424.

13. Wang, X.; Guan, R.G.; Misra, R.D.K.; Wang, Y.; Li, H.C.; Shang, Y.Q. The mechanistic contribution of nanosized Al_3Fe phase on the mechanical properties of Al-Fe alloy. *Mater. Sci. Eng. A* **2018**, *724*, 452–460. [CrossRef]

14. Belov, N.A.; Aksenov, A.A.; Eskin, D.G. *Iron in Aluminum Alloys: Impurity and Alloying Element*; CRC Press: Boca Raton, FL, USA, 2002; p. 360.

15. Santora, E.; Berneder, J.; Simetsberger, F.; Doberer, M. Mechanical properties evolution for 8xxx foil stock materials by alloy optimization—Literature review and experimental research. In *Light Metals 2019*; Springer: Cham, Switzerland, 2019; pp. 365–372.

16. Li, Y.J.; Arnberg, L. A eutectoid phase transformation for the primary intermetallic particle from $Al_m(Fe,Mn)$ to $Al_3(Fe,Mn)$ in AA5182 alloy. *Acta Mater.* **2004**, *52*, 2945–2952. [CrossRef]

17. Slámová, M.; Karlík, M.; Cieslar, M.; Chalupa, B.; Merle, P.; Structure transformations during annealing of twin-roll cast Al-Fe-Mn-Si (AA 8006) alloy sheets I. Effect of cold rolling and heating rate. *Kovové Materiály* **2002**, *40*, 389–401.

18. Stanica, C.; Moldovan, P.; Dobra, G.; Bojin, D.; Stanescu, C. Characterization of the microstructure of the AA8006 alloy sheets and foils. In *Light Metals 2008*; TMS (The Minerals, Metals & Materials Society): New Orleans, LA, USA, 2008; pp. 703–707.

19. Băjenaru, O.; Moldovan, P.; Bojin, D. Defects in the AA8006 aluminum alloy sheets. *UPB Sci. Bull. Ser. B Chem. Mater. Sci.* **2008**, *70*, 103–108.

20. Moldovan, P.; Popescu, G.; Miculescu, F. Microscopic study regarding the microstructure evolution of the 8006 alloy in the plastic deformation process. *J. Mater. Process. Technol.* **2004**, *153*, 408–415. [CrossRef]

21. Chen, Z.; Li, S.; Zhao, J. Homogenization of twin-roll cast A8006 alloy. *Trans. Nonferrous Metal Soc. Chin.* **2012**, *22*, 1280–1285. [CrossRef]

22. Sadeghi, I.; Wells, M.A.; Esmaeili, S. Modeling homogenization behavior of Al-Si-Cu-Mg aluminum alloy. *Mater. Des.* **2017**, *128*, 241–249. [CrossRef]

23. Allena, C.M.; O'Reilly, K.A.Q.; Cantor, B.; Evans, P.V. Intermetallic phase selection in 1XXX Al alloys. *Prog. Mater. Sci.* **1998**, *43*, 89–170. [CrossRef]

Article

Computational Fatigue Analysis of Auxetic Cellular Structures Made of SLM AlSi10Mg Alloy

Miran Ulbin, Matej Borovinšek, Matej Vesenjak and Srečko Glodež *

University of Maribor, Faculty of Mechanical Engineering, Smetanova 17, 2000 Maribor, Slovenia;
miran.ulbin@um.si (M.U.); matej.borovinsek@um.si (M.B.); matej.vesenjak@um.si (M.V.)
* Correspondence: srecko.glodez@um.si; Tel.: +386-2-2207708

Received: 26 May 2020; Accepted: 10 July 2020; Published: 14 July 2020

Abstract: In this study, a computational fatigue analysis of topology optimised auxetic cellular structures made of Selective Laser Melting (SLM) AlSi10Mg alloy is presented. Structures were selected from the Pareto front obtained by the multi-objective optimisation. Five structures with different negative Poisson's ratios were considered for the parametric numerical analysis, where the fillet radius of cellular struts has been chosen as a parameter. The fatigue life of the analysed structures was determined by the strain–life approach using the Universal Slope method, where the needed material parameters were determined according to the experimental results obtained by quasi-static unidirectional tensile tests. The obtained computational results have shown that generally less auxetic structures tend to have a better fatigue life expectancy. Furthermore, the fillet radius has a significant impact on fatigue life. In general, the fatigue life decreases for smaller fillet radiuses (less than 0.3 mm) as a consequence of the high-stress concentrations, and also for larger fillet radiuses (more than 0.6 mm) due to the moving of the plastic zone away from the edge of the cell connections. The obtained computational results serve as a basis for further investigation, which should be focused on the experimental testing of the fabricated auxetic cellular structures made of SLM AlSi10Mg alloy under cyclic loading conditions.

Keywords: aluminium alloy; SLM; auxetic structures; numerical analysis; fatigue

---

1. Introduction

Additive Manufacturing (AM) is a fabrication process that provides unique opportunities to manufacture customised parts with complex geometries or functionally graded materials [1,2]. As presented in [3,4], AM brings the promise of a new industrial revolution, offering the capability to manufacture parts through the repetitive deposition of material layers directly from a digital Computer-Aided Design (CAD) model. For those reasons, AM applies to a wide range of industries [5,6], including medical equipment [7,8], automotive [9,10], aerospace [11,12], and other consumer products [13]. Many AM technologies are being currently used in different engineering applications [14,15], e.g., Direct Metal Laser Sintering (DMLS), Selective Laser Melting (SLM), Electron Beam Melting (EBM), and Laser Metal Deposition (LMD). Among the AM processes, DMLS and SLM techniques are particularly widespread, especially for the aluminium alloy processing [16,17]. The DMLS technology is appropriate to build parts out of any metal alloy, while SLM can only be used with certain metals.

For AM technologies, Al–Si or Al–Si–Mg are the common alloys. In general, the static strength of AM parts depends on the density of the parts, as well as on the microstructure formed during the fabrication. As presented by Aboulkhair et al. [18] and Kempen et al. [19], the microstructure of the AM fabricated parts made of the AlSi10Mg alloy is usually finer (higher static strength) compared to the parts, which are fabricated via classical procedures (e.g., casting). On the other hand, the microstructure of the AM fabricated parts is generally anisotropic and depends on the building and scanning directions during the AM process. For that reason, the monotonic material properties (yield stress, ultimate tensile strength) are also anisotropic and may strongly depend on the orientation of texture [20–22]. Zhou et al. [23] have shown that the microstructure, hardness, and static strength of the AM parts may be significantly influenced by additional heat treatment. Nevertheless, the fatigue behaviour of the AM parts is not well understood yet, although a lot of research was performed in the last decade [24]. Similar to the static mechanical properties, the fatigue strength of the AM parts primarily depends on their microstructure. However, the AM process usually leads to an increased surface roughness of the fabricated part, which decreases its fatigue strength [25]. Furthermore, material defects such as the porosity and insufficient layer bonding could result in an increased scatter of the fatigue properties [26]. Besides the fatigue analyses, the investigations on the fracture behaviour of the AM components have also been carried out, especially due to the high importance in the automotive industry and aerospace applications [27–29].

As mentioned above, AM technologies, including the SLM process, provide opportunities to manufacture very complex geometries or functionally graded materials, which are inaccessible through traditional manufacturing techniques. Typical lightweight structures, where the SLM process may be used, are pre-designed advanced cellular metamaterials and structures, which are difficult to fabricate by the traditional processes. In the last period, the research has been focusing on the mechanical response of the gradient lattice structures manufactured via the AM process in terms of energy absorption under the compressive loading conditions [30–32]. Some authors also investigated the fatigue behaviour of different AM lattice structures considering the effect of the cellular topology and geometrical imperfections on the fatigue life of the analysed structures [33–35]. The next group of metamaterials suitable to be produced using the AM technology are the auxetic cellular structures, which are the subject of this study (for a detailed description of the auxetic cellular structures, see Section 2.1).

The research work in the presented study is focused on the fatigue behaviour of five different auxetic cellular structures obtained by the topology optimisation and made of the SLM AlSi10Mg alloy, which is widely used in the aerospace and automotive industry because of its high specific strength, high corrosion resistance, and good flowability [36,37]. The fatigue life of different auxetic cellular geometries, also considering the influence of the fillet radiuses, is determined by the strain–life approach using Universal Slope method [38]. The material parameters were determined according to the experimental results previously obtained by the unidirectional quasi-static tests [39]. Numerical nonlinear finite element simulations were performed to determine the deformation behaviour and the stress and strain fields in the critical cross-sections of the analysed cellular structures. Then, the numerical results were used as the input parameters in the subsequent fatigue analyses.

2. Selection of Optimised Auxetic Cellular Structures

2.1. Auxetic Cellular Structures

Cellular structures have been increasingly used in modern engineering applications over the past decades due to their attractive combination of mechanical and thermal properties [40–42]. Recent advances in the AM technologies enable the fabrication of parts with complex internal cellular structure, which are commonly adjusted for specific applications. The auxetic cellular structures exhibit a negative Poisson's ratio and thus experience large volume changes under loading [43–45]. The auxetic structures tend to expand in the lateral direction when subjected to the uniaxial tensile loading and vice versa in the case of the compressive loading (Figure 1).

Such behaviour is made possible by their 2D or 3D designed hinge-like cellular skeleton with predefined geometry [46–48]. The auxetic cellular structures (made of metals [49], polymers [50,51], and textiles [52–54]) offer extraordinary mechanical properties (under quasi-static [55], impact [56,57], and fatigue [58,59] loading), i.e., variable stiffness [60], low density, large mechanical energy absorption through deformation [61–63], and unique deformation behaviour [64,65]. Their cell shape changes rapidly through the complete auxetic structure under loading; thus, the loading efficiently spreads through the structure. The impact on a small part of the auxetic structure results in energy dissipation through its entire structure. With recent advances in the AM technologies [66], it is possible to fabricate the auxetic cellular structures with optimised and exactly predefined geometries (i.e., strut thickness, geometry, cell shape and size). The design and optimisation of auxetic structures can be done by using advanced Computer-Aided Engineering (CAE) methodologies to achieve and control the desired mechanical properties for specific applications [60,67]. Such precise fabrication control of the shape, size, and distribution of cells with the use of layered additive technologies makes the auxetic structures superior to the other conventionally manufactured cellular materials.

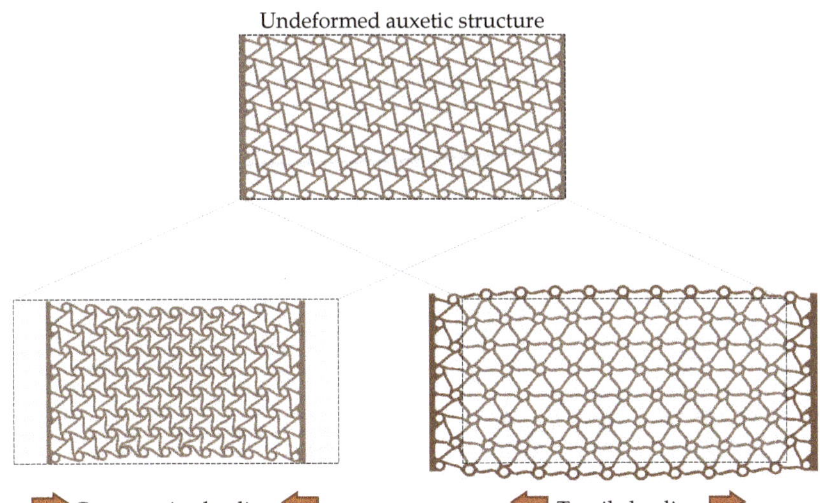

Figure 1. The behaviour of auxetic cellular structures under compressive and tensile loading.

2.2. Topology Optimised Auxetic Structures

Many auxetic cellular structures have been developed up to date [45]. Their shape, topology, and dimensions were usually custom-made for the task for which they were designed. Therefore, it is difficult to compare the fatigue properties between them. In the previous study by Borovinšek et al. [67], a procedure for the determination of optimal auxetic topology was developed using a numerical approach. The auxetic topology was modelled as a quarter cell of a 2D planar periodic structure. Appropriate boundary conditions were applied to account for the rest of the structure. The quarter cell model had a square shape, and its topology was represented by a grid of 10×10 equal squares (parameter regions), each of which could either represent material or a void region (Figure 2a). In this way, any topology of the quarter cell could be obtained by changing the content of the parameter regions.

The performance of the auxetic structure was regarded as a performance of the compliant mechanism, and as such, two objective functions were applied to determine the optimal topologies. The first objective function was used to measure the stiffness of the auxetic structure, while the second objective function measured its flexibility. Using two objective functions and a multi-objective optimisation procedure, which changed the values of the parameter regions, resulted in a set of best topologies called the Pareto front [67]. All solutions on the Pareto front are optimal and equally good. Thus, applying this procedure results in a set of topologies, which have an auxetic behaviour. Since they are optimised with the same procedure and with the same objectives, it is also possible to directly compare their fatigue behaviour.

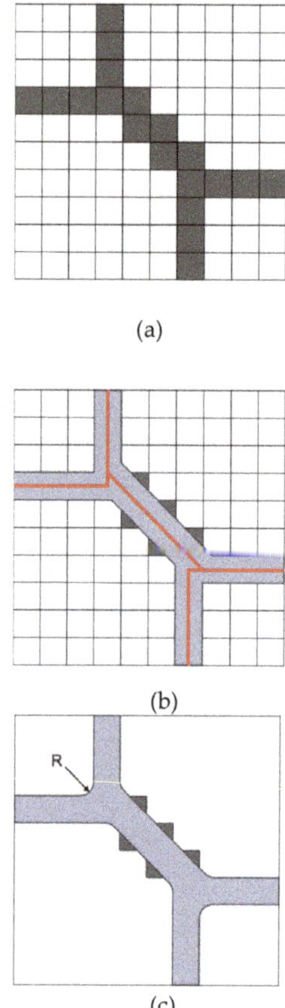

Figure 2. Base topology (**a**); medial axis topology (**b**); and rounded medial axis topology (c).

The drawback of this procedure is that the resulting optimal topologies of the quarter cell models are sharp-edged, since their topology consists of square-shaped parameter regions (Figure 2a). Sharp edges of the structure result in stress concentrations under load and do not allow a direct application of the obtained topologies from [67]. To study the fatigue behaviour of the obtained topologies, a geometry smoothing procedure was developed for this investigation. It was applied to reduce the number of possible stress concentration regions.

The smoothing procedure consists of the following five steps:

1. Selection of the topology from the Pareto front (Figure 2a, dark grey area).
2. Determination of the medial axis of the topology.
3. Simplification of the medial axis to straight lines between the line connection points (Figure 2b, red lines).
4. Thickening of the simplified medial axis to obtain the geometry; the horizontal and vertical lines retained their thickness, while a new thickness was defined to the oblique lines of the medial axis. The thickness of these lines was set to such a value that the topology area remained the same (Figure 2b, light grey area).
5. All remaining sharp corners were rounded using an equal fillet radius R (Figure 2c).

From the Pareto front, obtained by the multi-objective optimisation, which contained 200 different topologies, five base topologies were selected for this study. The topologies of the Pareto front were first sorted by their Poisson's ratio starting with the lowest one (Topology 1), followed by the median one (Topology 3), and so on until the highest one (Topology 5). An additional two topologies were selected between the lowest and the median Poisson's ratio (Topology 2) and between the median and the highest Poisson's ratio (Topology 4). The resulting five topologies are almost equally apart from each other on the Pareto front; thus, they are a good representation of the best solution set. The Poisson's ratio and the mass of all five base topologies are reported in Chapter 3. After selection, the smoothing procedure was applied, which resulted in five distinct quarter unit cell topologies (Figure 3). Their complete auxetic structures were made by mirroring and multiplying the quarter cell topology in such a way that the resulting periodic structures consisted of six-unit cells in both the (x and y) coordinate directions.

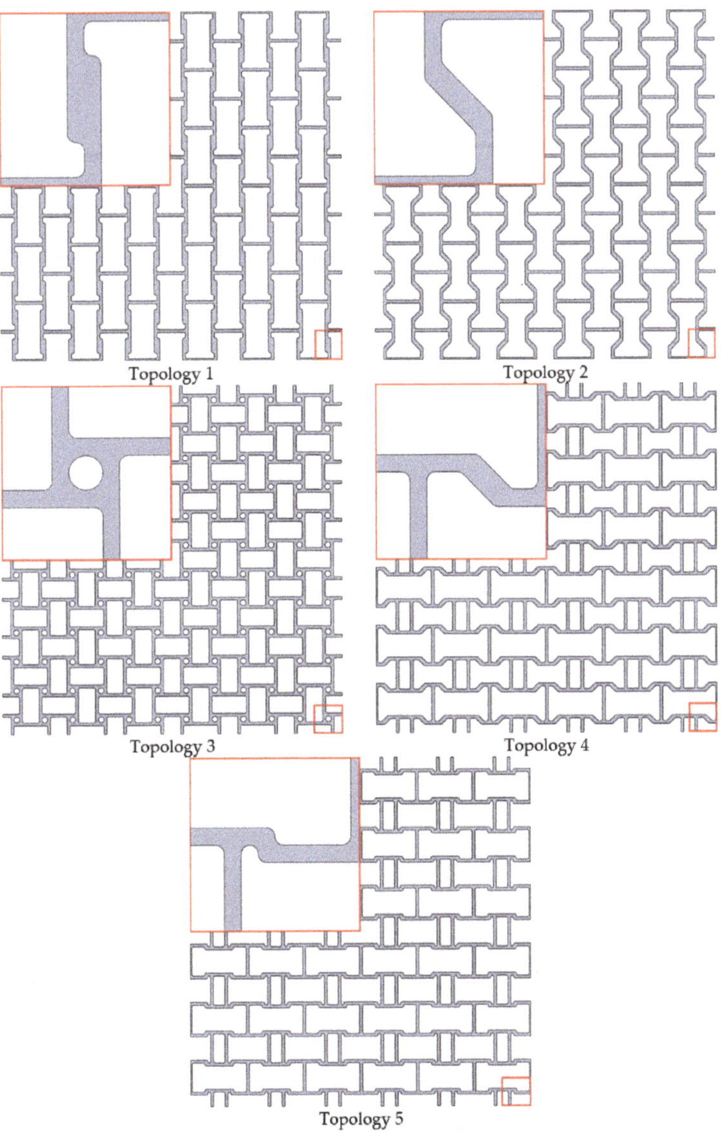

Figure 3. Selected quarter unit cell topologies and resulting auxetic structures.

3. Material Characterisation and Fatigue Analyses

3.1. Material Characterisation

The material used in this study is aluminium alloy AlSi10Mg. The specimens used for the further tensile tests have been fabricated by the SLM technology using the EOSINT-M-270 system without additional polishing. The general process data and the material composition are shown in Table 1. Figure 4a shows the geometry of the flat specimen, while the designated building and scanning fabrication directions are shown in Figure 4b.

Table 1. Material data sheet of Selective Laser Melting (SLM) AlSi10Mg alloy.

Layer thickness: 30 μm										
Volume rate with the standard parameters: 4.8 mm^2/s										
Surface roughness (as-built, cleaned): *Ra* 15–19 μm; *Rz* 96–115 μm										
The density with the standard parameters: 2.68 g/cm^3										
Material composition (maximum values in %)										
Al	Si	Fe	Cu	Mn	Mg	Ni	Zn	Pb	Sn	Ti
Bal.	9–11	0.55	0.05	0.45	0.2–0.45	0.05	0.1	0.05	0.05	0.15

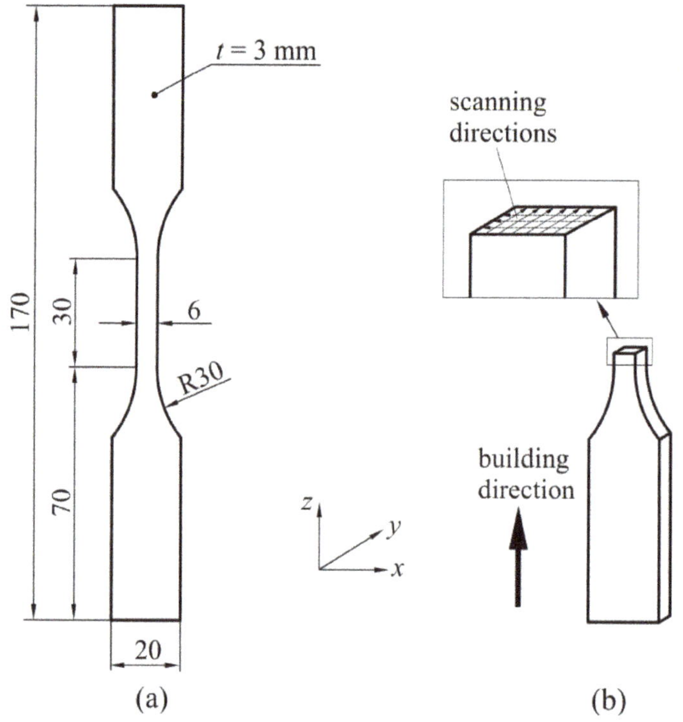

Figure 4. Geometry (a) and building direction (b) of a flat specimen. (all dimensions are in millimetres [mm]).

Figure 5 shows the polished surface of the fabricated specimen. The black circles represent the pores. They are significantly more frequent and larger near the specimen surface all around the circumference, up to 500 μm inwards. These near-surface pores were formed at the beginning/end of a scan in the *x*-direction and were not removed by the laser movement in the *y*-direction during the formation of the subsequent layer.

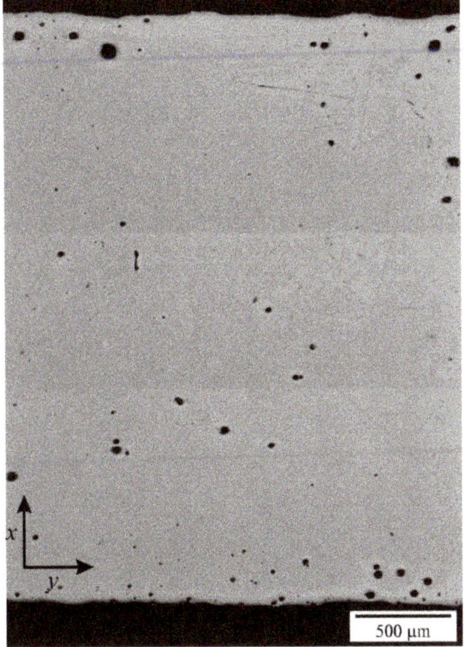

Figure 5. A light micrograph of the polished specimen [39].

Two specimens were used for the quasi-static tensile tests to measure the stress–strain (σ–ε) response. The tensile tests were carried out using the 100 kN MTS Landmark hydraulic machine (MTS Systems GmbH, Germany) at a room temperature of 23 °C. The tests were displacement controlled with a loading rate of 0.5 mm/min according to the DIN-ISO 6892 standard. The force was measured with the 100 kN MTS load cell, and the strains were measured with the MTS 834.11F-24 extensometer (MTS Systems GmbH, Germany). The measured quasi-static stress–strain responses [39] were almost the same for both specimens and are in a good agreement if compared to the experimental results from other researchers [68]. Based on the measured stress–strain responses (Figure 6), the average material parameters have been determined (Table 2).

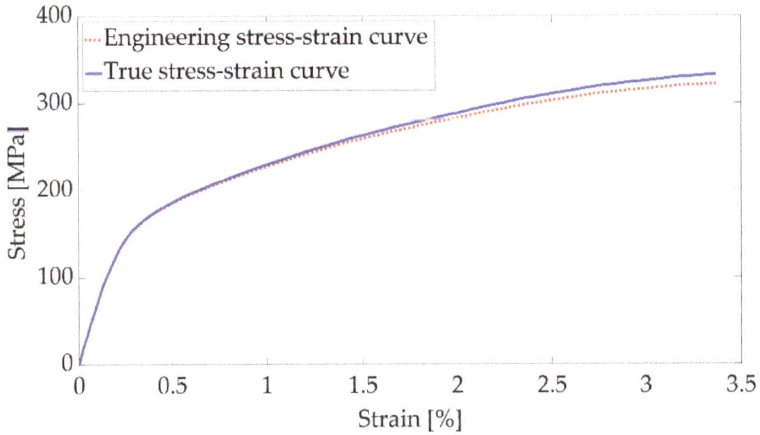

Figure 6. Stress–strain response of the SLM AlSi10Mg alloy.

Table 2. Material parameters of the SLM AlSi10Mg alloy.

Elastic Modulus E (MPa)	Poisson's Ratio ν (-)	Yield Stress $R_{p0,2}$ (MPa)	Ultimate Tensile Strength UTS R_m (MPa)	Strain at Rupture A_5 (%)	True Fracture Strain ε_f (-)
70,900	0.3	180	318	3.35	0.031

The true stress–strain (σ_{true}–ε_{true}) data were determined using engineering stress (σ_{eng}) and engineering strain (ε_{eng}) values as follows:

$$\varepsilon_{true} = \ln\left(1 + \varepsilon_{eng}\right) \tag{1}$$

$$\sigma_{true} = \sigma_{eng} \times \left(1 + \varepsilon_{eng}\right). \tag{2}$$

3.2. Strength Analysis

The numerical models of five auxetic structures were built in Ansys Mechanical [69], using a 2D plane stress formulation. The size of 10 mm × 10 mm was chosen for the one-unit cell, which results in a structure size of 60 mm × 60 mm and a nominal strut thickness of 0.5 mm. The finite element mesh consisted of plane stress parabolic quad finite elements PLANE183 with eight nodes. With the parametrical analysis, it was deduced that the results converge using a mesh, where the global element size is approximately 0.2 mm, with a consideration of curvatures, where the element size is even smaller. This resulted in numerical models with an average of 30,000 finite elements and 100,000 nodes. Each topology was constrained in the y-direction at the bottom and loaded with displacement at the top. All five structures were loaded with the same displacement of 0.4 mm in the + y-direction, thus creating a tension in the vertical direction. Such loading was chosen to induce plasticity in all structures. The higher loading displacement would cause severe plasticity in some structures. Similarly, a lower value of displacement would not cause plastic strain in some structures. The models were constrained in the x-direction in the middle of the structure to prevent horizontal movement. The boundary conditions are presented in Figure 7, where loading is acting in the vertical direction (y-direction).

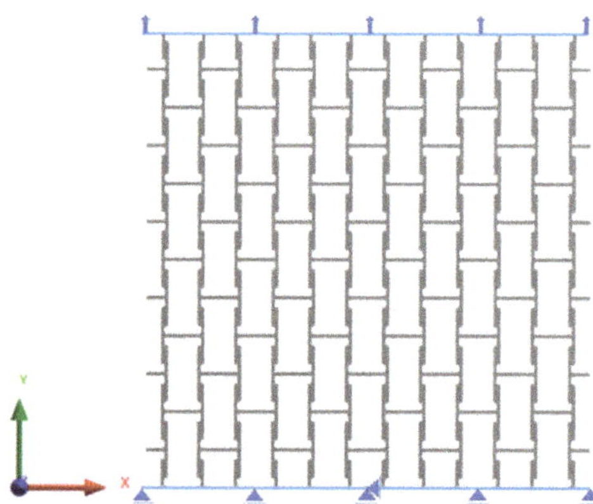

Figure 7. Boundary conditions.

The material in all presented numerical models was the SLM AlSi10Mg alloy with the properties presented in Section 3.1. The models were loaded with the displacement, which caused plastic deformation in all five models. Therefore, material nonlinearity was considered in the simulations. For plasticity, a multilinear isotropic hardening model was used, with the utilisation of the true stress–strain material data shown in Figure 6.

Numerical results in the form of total displacements are presented in Figure 8. The resulting displacement was scaled by the factor 11 for a better visualisation of all structures. It can be noted that they have different Poisson ratios. The Poisson ratios presented in Table 3 were calculated as the ratio between the longitudinal and lateral displacement. The Poisson ratios of the analysed structures are, to some extent, different than the unit cell's Poisson ratios. All the calculated Poisson ratios are negative, which proves that all the analysed structures are also auxetic. It can be seen that Topology 1 has the lowest Poisson ratio, while Topology 5 has the highest Poisson ratio. The Poisson ratios of the other three topologies are ascending between Topology 1 and Topology 5. Table 3 also reports the mass of each model showing that all structures have a similar weight. Only Topology 3 is a bit more massive.

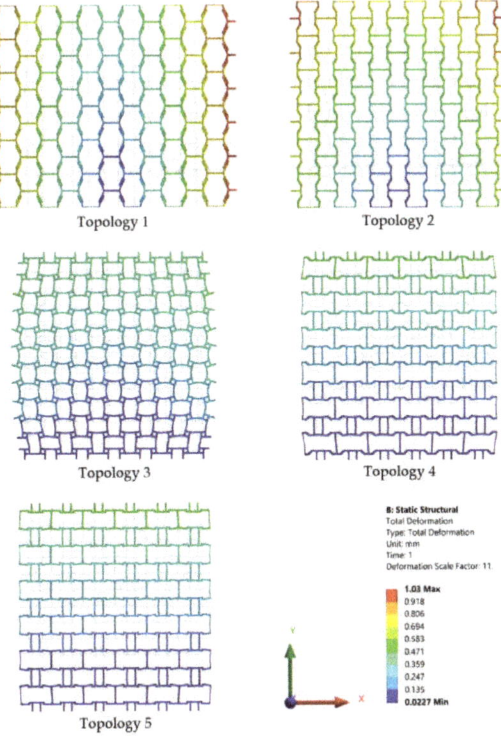

Figure 8. Deformed and undeformed structures.

Table 3. Properties and results of the analysed topologies.

Topology	Mass (g)	Displacement x-direction (mm)	Equivalent Elastic Strain (mm/mm)	Equivalent Plastic Strain (mm/mm)	Poisson Ratio (-)	Unit Cell's Poisson Ratio (-)
1	0.10	0.964	0.00408	0.0154	−2.41	−4.01
2	0.09	0.381	0.00307	0.0180	−0.95	−2.71
3	0.12	0.280	0.00362	0.0094	−0.67	−0.92
4	0.10	0.080	0.00312	0.0022	−0.20	−0.35
5	0.09	0.040	0.00323	0.0011	−0.01	−0.19

Figure 9 shows the location of the maximum plastic strain for each model. The area of the maximum plastic strain is also magnified to visualise the exact location of the developed plastic zones. Location of the maximum plastic strain is in all cases at the fillet radius of the auxetic structure.

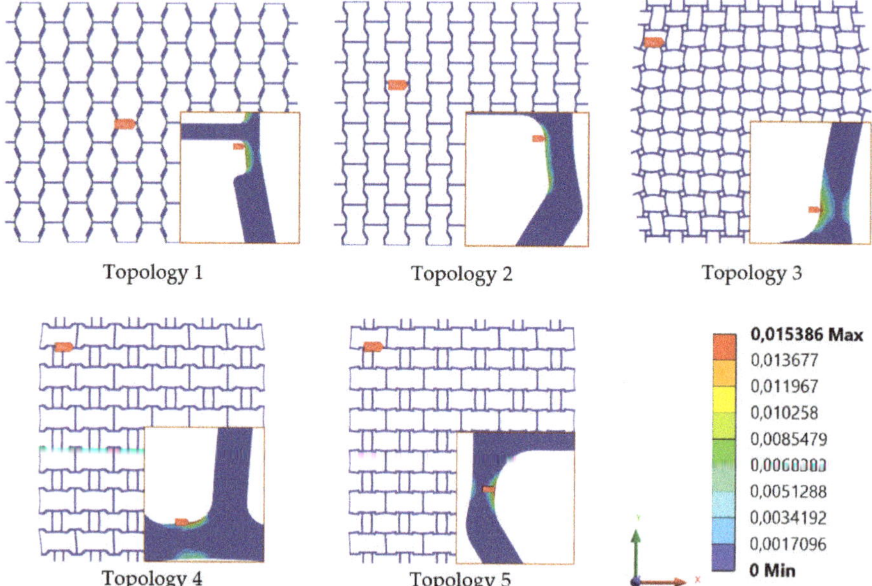

Figure 9. Location of the maximum equivalent plastic strain and its distribution.

Since the size of the radius is important, a series of parametric analyses were conducted with different radius sizes. Radiuses were scaled in a range of 0.1 and 0.9 mm with a step of 0.1 mm. Figure 3 shows the basic topology designs from which the numerical models were built. Most of the structures allow a wide range of radiuses, while Topology 1 and Topology 5 are limited because the two fillets intersect and prevent further radius scaling. Therefore, some radiuses were scaled, while the potentially intersecting radiuses were presented only up to the maximum possible values, e.g., Topologies 1 and 5. Each time the radius was scaled, the numerical model was regenerated with the same mesh parameters and the same boundary conditions as presented in Figure 7. In that way, it was possible to determine how radius sensitive the analysed structures are and the optimal radius for each structure in terms of fatigue.

Figure 10 shows the equivalent plastic strain for different fillet radiuses. For Topology 1, some fillets could not be scaled beyond 0.25 mm, while the remaining fillets are limited to 0.75 mm; therefore, the results for the radius of 0.9 mm for Topology 1 are missing.

From the results in Figure 10, it is clear that at a small fillet radius, the plastic strain occurs near the edge, which is placed where the crack initialisation is expected. With a larger fillet radius, the area of the plastic zone is wider, and the value of the equivalent plastic strain is lower in all structures. In most of the observed structures, this occurs at the radius of approximately 0.4 mm, with exceptions in Topology 1 (at the radius of approximately 0.2 mm) and Topology 4 (at the radius of approximately 0.3 mm). With further increase of the fillet radius, the value of the equivalent plastic strain increases. Furthermore, some new critical areas occur. Accordingly, large fillet radiuses create new critical places for the crack initiation, where sometimes the equivalent plastic strain has the same or larger value than observed in the case of very small fillet radiuses.

Then, the obtained numerical results were used for fatigue life determination. From each numerical result, the maximum equivalent elastic strain was added to the maximum equivalent plastic strain. This allows the calculation of the fatigue life cycle for each topology, including the fillet radius variations (Figure 10). For each structure, additional numerical analysis was performed to determine the yield load condition, when the structure behaves purely elastic.

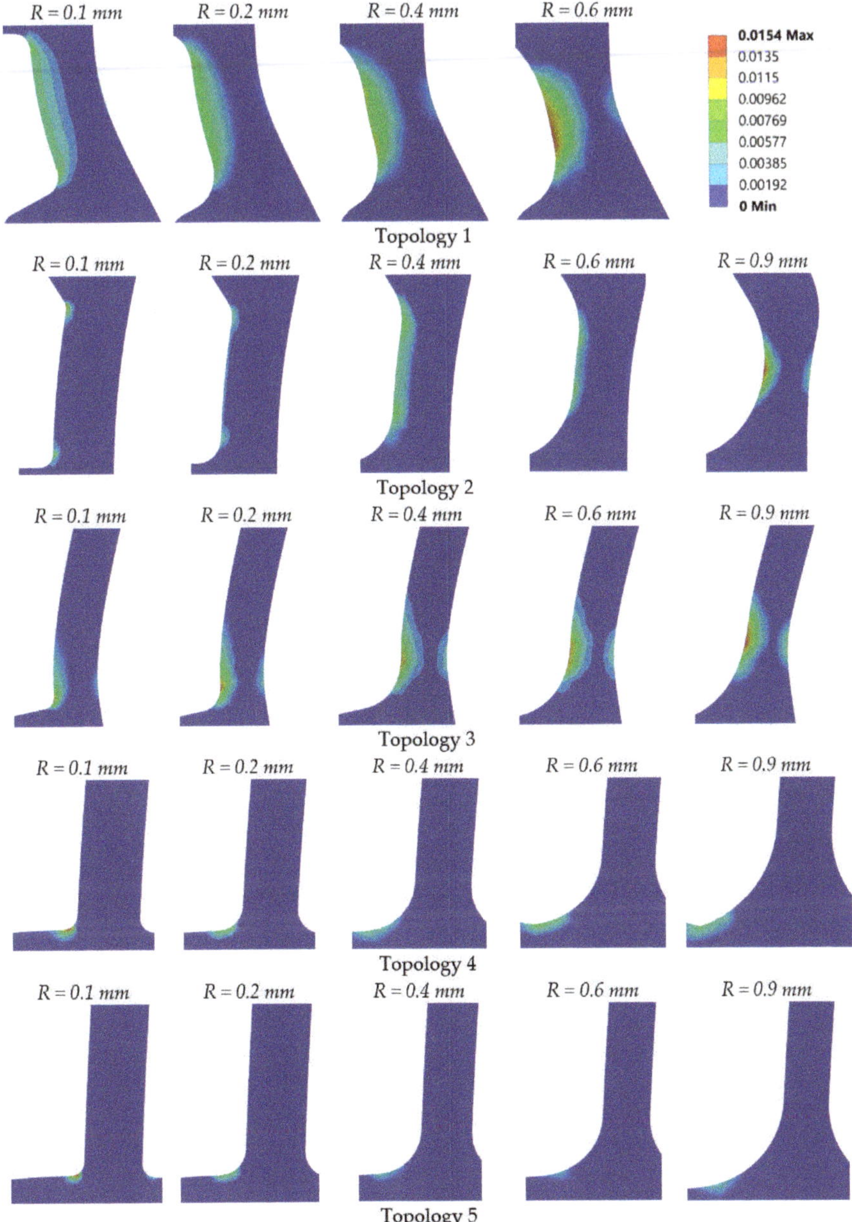

Figure 10. Equivalent plastic strain for different fillet radiuses.

3.3. Fatigue Life Determination

In the presented study, the strain–life approach is used to determine the fatigue life of treated cellular structures. In general, this approach is based on the knowledge of the stresses and strains that occur at locations, where the fatigue crack nucleation is likely to start, e.g., holes, fillets, grooves. In this approach, the fatigue behaviour of the material is described by the strain–life curve expressed by the Coffin–Manson relation between fatigue life N_i and the total amplitude strain [70]. However, the material low-cycle fatigue parameters should be determined previously using the appropriate low-cycle fatigue test. If it is not the case, an alternative method based on the material parameters obtained from a quasi-static test can be used. In this study, the method of Universal Slopes as proposed by Muralidharan and Manson [38] has been used to obtain the number of loading cycles N_i required for the fatigue crack initiation in the critical cross-section of the treated cellular structure:

$$\varepsilon_{Ea} = 0.623\left(\frac{R_m}{E}\right)^{0.832}(2N_i)^{-0.09} + 0.0196(\varepsilon_f)^{0.155}\left(\frac{R_m}{E}\right)^{-0.53}(2N_i)^{-0.56} \tag{3}$$

where ε_{Ea} is the equivalent amplitude strain, E is the elasticity modulus, R_m is the ultimate tensile strength, and ε_f is the true fracture strain. Once the equivalent amplitude strain ε_{Ea} is known, the number of stress cycles N_i can be obtained iteratively from Equation (3) using Matlab software [71] and with a consideration of the material parameters E, R_m, and ε_f given in Table 2. The results are shown in Figure 11.

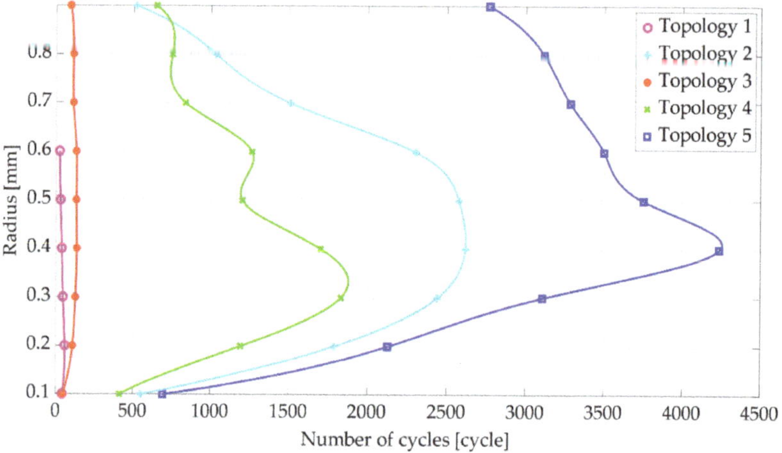

Figure 11. The fatigue life N_i of different topologies and fillet radiuses.

For Topology 1, the expected fatigue life cycles are rather low, with a maximum value at the fillet radius of 0.2 mm. Topology 4 resulted in the best fatigue life expectance with the fillet radius of 0.3 mm. All other structures have the highest number of cycles with a radius of 0.4 mm, while the life expectancy of Topology 3 is very low. The best fatigue life is expected for Topology 5, which also has the highest Poisson ratio, making it the least auxetic structure presented in the paper. On the other hand, Topology 2 has the second-highest fatigue life expectancy, while having the second-lowest Poisson ratio. Therefore, Topology 2 could be used in cases where a highly auxetic and durable structure is needed.

4. Discussion

Five cellular structures were considered for the numerical analysis to identify the most suitable auxetic topology when considering the fatigue life under tension loading conditions. Structures were selected from the Pareto front obtained by the multi-objective optimisation. The structures were sorted by their Poisson ratio, starting with the lowest one (Topology 1) to the one that has the Poisson's ratio only slightly below zero (Topology 5), as can be seen from Table 3.

All five topologies were subjected to the same tensile loading conditions. The amount of the tensile load was high enough to cause plasticity in every topology but small enough that each topology would sustain several loading cycles until failure. The presented results of the numerical analysis show that generally less auxetic structures tend to have a better fatigue life expectancy.

The fillet radius has a significant impact on fatigue life. While it is expected that smaller radiuses would increase the probability of crack initiation, it is interesting to note that larger fillet radiuses also negatively influence the fatigue life. As the fillet radius increases, the plastic zone is moved away from the edge, as can be seen from Figure 10. Global tension loading displacement induces hinge-like behaviour in structure corners. While the hinges rotate, the structures struts are primarily loaded in bending. Since the struts have a constant cross-section area, the internal bending moment is equally distributed along their length, and the whole strut deforms to account for the hinge rotation. With increased fillet radius, the strut length is decreased. To account for the same hinge rotation, a shorter strut must deform more, which causes a higher rate of plasticity with the same tension loading displacement.

Topology 1 has the lowest (most negative) Poisson ratio, which makes it the most auxetic structure analysed herein. Computed displacements in the lateral direction (x-direction) are more than two times larger than the loading displacements in the longitudinal direction (y-direction). However, the prescribed tension displacement is quite large for this structure, as it results in an equivalent plastic strain of 1.5%. The fatigue life under such loading conditions would be very short, which is also clearly shown in Figure 11. Loading would have to be reduced at least 10 times to achieve the pure elastic behaviour of the structure, which questions its usability compared to the other analysed structures.

As can be seen from Figure 11, Topology 2 is the second most auxetic structure and has the second-best fatigue life. This makes Topology 2 the best choice for an application where the auxetic response is needed. With the Poisson ratio of −0.95, the displacement in the lateral direction (x-direction) is almost identical to the tension loading displacement. With the applied loading, the structure would sustain approximately 2500 cycles, and if the loading displacement would be decreased to less than half of the applied loading, the structure would behave purely elastically, while still exhibiting the same displacement in both directions. This makes Topology 2 the most appropriate choice for the auxetic structure.

By analysing Topology 3, it can be observed that the lateral displacement is nearly half the size of the tension loading displacement. However, the equivalent plastic strain is approximately 1%, and the structure has the second-worst fatigue life expectancy Additionally, Topology 3 is the most massive structure compared to the other topologies. Therefore, this structure is not recommended for an application subjected to cyclic loading unless a specific need utilising such a Poisson ratio is required.

The Poisson ratio of Topology 4 is equal to −0.2. Topology 4 also has a high expected fatigue life. Similar to Topology 2, this structure would behave purely elastically when the loading displacement is reduced by half. Comparing Poisson ratio, it is evident that the lateral displacement would be five times lower, as it was observed in the case of Topology 2. It could be recommended for applications where a small amount of auxetic behaviour is required. Nevertheless, Topology 2 shows a better performance in terms of lateral displacement and fatigue life.

Topology 5 shows the smallest auxetic effect, as its Poisson ratio is close to zero. At the same time, it is the structure with the highest number of fatigue life cycles determined, as can be seen from Figure 11. If the tension loading displacement would be reduced only by 25%, the structure would behave purely elastic. Therefore, Topology 5 could be used for applications where the transversal strain should be limited.

5. Conclusions

In this study, the fatigue behaviour of auxetic cellular structures obtained by multi-objective topology optimisation and made of SLM AlSi10Mg alloy is presented. Five structures with different Poisson ratios were considered for the parametric numerical analysis, where the fillet radius of cellular struts has been chosen as a parameter. The fatigue life of the analysed structures was determined by the strain–life approach using the Universal Slope method, where the needed material parameters were determined according to the experimental results obtained by the quasi-static unidirectional tensile tests. Based on the fractography of the quasi-static specimens, numerical simulations, and fatigue analyses, the following conclusions can be made:

- The microstructure of treated SLM AlSi10Mg alloy consists of bands, which were produced by varying the scan direction in each subsequent layer. The material porosity is more frequent and larger in the near-surface layers, where pores were formed at the end/beginning of scanning during the SLM process.
- The obtained material parameters by the quasi-static tests (proportional limit, ultimate tensile strength, strain at rupture) are comparable to the results presented by the other researchers.
- The obtained computational results have shown that less auxetic structures (higher Poisson ratios) tend to have a better fatigue life expectancy (the longest fatigue life has been obtained for the structure topology with the minimum auxetic characteristics).
- The fillet radius of cellular struts has a significant impact on fatigue life. Computational analyses have shown that the fatigue life decreases for smaller fillet radiuses (less than 0.3 mm) as a consequence of high-stress concentrations and also for larger fillet radiuses (more than 0.6 mm) due to the movement of the plastic zone away from the edge of the cell connections. Besides the fillet radius sizes, other reasons for stress concentrations might exist. This should be investigated in future research.
- The fatigue life in this study was obtained using the simplified Universal Slope method, where the material parameters were determined by quasi-static tensile tests considering the partly porous structure of analysed SLM AlSi10Mg alloy (Figure 5). For a more accurate determination of fatigue life of real components made of SLM AlSi10Mg alloy, the comprehensive Low Cycle Fatigue (LCF) tests should be performed to determine the appropriate LCF material properties, where the influence of pores will be considered in detail. Once the LCF material parameters are known, the standardised strain life approach could be used for the subsequent fatigue analysis.
- The obtained computational results provide a basis for further investigation including experimental testing of the fabricated auxetic cellular structures made of SLM AlSi10Mg alloy under the cyclic loading conditions.

Author Contributions: Conceptualization, M.U. and S.G.; investigation, S.G.; methodology, M.V.; project administration, M.V.; software, M.B.; Writing—review and editing, M.U., M.B., M.V. and S.G. All authors have read and agreed to the published version of the manuscript.

Funding: This research was funded by the Slovenian Research Agency—ARRS, basic research project No. J2-8186 and research core No. P2-0063.

Conflicts of Interest: The authors declare no conflict of interest.

References

1. Savolainen, J.; Collan, M. How additive manufacturing technology changes business models? Review of literature. *Addit. Manuf.* **2020**, *32*, 101070. [CrossRef]
2. Plocher, J.; Panesar, A. Review on design and structural optimisation in additive manufacturing: Towards next-generation lightweight structures. *Mater. Des.* **2019**, *183*, 108164. [CrossRef]
3. Chunyong, L.; Yazhou, H.; Ning, L.; Xianrui, Z.; Hongshui, W.; Xinping, Z.; Yulan, F.; Jingyun, H. Laser polishing of Ti6Al4V fabricated by selective laser melting. *Metals* **2020**, *10*, 191. [CrossRef]
4. Mugwagwa, L.; Yadroitsev, I.; Matope, S. Effect of process parameters on residual stresses, distortions, and porosity in selective laser melting of maraging steel 300. *Metals* **2019**, *9*, 1042. [CrossRef]
5. Attaran, M. The rise of 3-D printing: The advantages of additive manufacturing over traditional manufacturing. *Bus. Horiz.* **2017**, *60*, 677. [CrossRef]
6. Gao, W.; Zhang, Y.; Ramanujan, D.; Ramani, K.; Chen, Y.; Williams, C.B.; Wang, C.L.; Shin, Y.C.; Zhang, S.; Zavattieri, P.D. The status, challenges, and future of additive manufacturing in engineering. *Comput. Aided Des.* **2015**, *69*, 65–89. [CrossRef]

7. Jardini, A.L.; Larosa, M.A.; Filho, R.M.; Zavaglia, C.A.C.; Bernardes, L.F.; Lambert, C.S.; Calderoni, D.R.; Kharmandayan, P. Cranial reconstruction: 3D biomodel and custom-built implant created using additive manufacturing. *J. Cranio-Maxillofac. Surg.* **2014**, *42*, 1877–1884. [CrossRef]

8. Mahmound, D.; Elbestawi, M. Lattice structures and functionally graded materials applications in additive manufacturing of orthopedic implants: A review. *J. Manuf. Mater. Process.* **2017**, *1*, 13. [CrossRef]

9. Chuang, C.H.; Chen, S.; Yang, R.J.; Vogiatzis, P. Topology optimisation with additive manufacturing consideration for vehicle load path development. *Int. J. Numer. Methods Eng.* **2017**, 1434–1445. [CrossRef]

10. Abdi, M.; Ashcroft, I.; Wildman, R.D. Design optimisation for an additively manufactured automotive component. *Int. J. Powertrains* **2018**, *7*, 142–161. [CrossRef]

11. Seabra, M.; Azevedo, J.; Araújo, A.; Reis, L.; Pinto, E.; Alves, N.; Santos, R.; Mortagua, J.P. Selective laser melting (SLM) and topology optimisation for lighter aerospace components. *Procedia Struct. Integr.* **2016**, *1*, 289–296. [CrossRef]

12. Zhu, J.H.; Zhang, W.H.; Xia, L. Topology optimisation in aircraft and aerospace structures design. *Arch. Comput. Methods Eng.* **2016**, *23*, 595–622. [CrossRef]

13. Bader, C.; Patrick, W.G.; Kolb, D.; Hays, S.G.; Keating, S.; Sharma, S.; Dikovsky, D.; Belocon, B.; Weaver, J.C.; Silver, P.A.; et al. Grown, printed, and biologically augmented: An additively manufactured microfluidic wearable, functionally templated for synthetic microbes, 3D print. *Arch. Addit. Manuf.* **2016**, *3*, 79–89. [CrossRef]

14. Simar, A.; Godet, S.; Watkons, T.R. Highlights of the special issue on metal additive manufacturing. *Mater. Charact.* **2018**, *143*, 1–4. [CrossRef]

15. Herzog, D.; Seyda, V.; Wycisk, E.; Emmelmann, C. Additive manufacturing of metals. *Acta Mater.* **2016**, *117*, 371–392. [CrossRef]

16. Ismael, A.; Wang, C.S. Effect of Nb additions on microstructure and properties of γ-TiAl based alloys fabricated by selective laser melting. *Trans. Nonferrous Metals Soc. China* **2019**, *29*, 1007–1016. [CrossRef]

17. Bagherifard, S.; Beretta, N.; Monti, S.; Riccio, M.; Bandini, M.; Guagliano, M. On the fatigue strength enhancement of additive manufactured AlSi10Mg parts by mechanical and thermal post-processing. *Mater. Des.* **2018**, *145*, 28–41. [CrossRef]

18. Aboulkhair, N.T.; Tuck, C.; Ashcroft, I.; Maskery, I.; Everitt, N.M. On the precipitation hardening of selective laser melted AlSi10Mg. *Metall. Mater. Trans. A* **2015**, *46*, 3337–3341. [CrossRef]

19. Kempen, K.; Thijs, L.; Van Humbeeck, J.; Kruth, J.P. Mechanical properties of AlSi10Mg produced by selective laser melting. *Phys. Procedia* **2012**, *39*, 439–446. [CrossRef]

20. Zhonghua, L.; Zezhou, K.; Peikang, B.; Yunfei, N.; Guang, F.; Wenpeng, L.; Shuai, Y. Microstructure and tensile properties of AlSi10Mg alloy manufactured by multi-laser beam selective laser melting (SLM). *Metals* **2019**, *9*, 1337. [CrossRef]

21. Fousováa, M.; Dvorskýa, D.; Michalcováa, A.; Vojtěcha, D. Changes in the microstructure and mechanical properties of additively manufactured AlSi10Mg alloy after exposure to elevated temperatures. *Mater. Charact.* **2018**, *137*, 119–126. [CrossRef]

22. Takata, N.; Kodaira, H.; Suzuki, A.; Kobashi, M. Size dependence of microstructure of AlSi10Mg alloy fabricated by selective laser melting. *Mater. Charact.* **2018**, *143*, 18–26. [CrossRef]

23. Zhou, L.; Mehta, A.; Schulz, E.; Mcwilliams, B.; Cho, K.; Sohn, Y. Microstructure, recipitates and hardness of selectively laser melted AlSi10Mg alloy before and after heat treatment. *Mater. Charact.* **2018**, *143*, 5–17. [CrossRef]

24. Daniewicz, S.R.; Shamsai, N. An introduction to the fatigue and fracture behaviour of additive manufactured parts. *Int. J. Fatigue* **2017**, *94*, 167. [CrossRef]

25. Beevers, E.; Brandão, A.D.; Gumpinger, J.; Gschweitl, M.; Seyfert, C.; Hofbauer, P.; Rohr, T.; Ghidini, T. Fatigue properties and material characteristics of additively manufactured AlSi10Mg—Effect of the contour parameter on the microstructure, density, residual stress, roughness and mechanical properties. *Int. J. Fatigue* **2018**, *117*, 148–162. [CrossRef]

26. Siddique, S.; Muhammad, I.; Wycisk, E.; Emmelmann, C.; Walther, F. Influence of process-induced microstructure and imperfections on mechanical properties of AlSi12 processed by selective laser melting. *J. Mater. Proc. Tech.* **2015**, *221*, 205–213. [CrossRef]

27. Siddique, S.; Awd, M.; Tenkamp, J.; Walther, F. Development of a stochastic approach for fatigue life prediction of AlSi12 alloy processed by selective laser melting. *Eng. Failure Anal.* **2017**, *79*, 34–50. [CrossRef]

28. Di Giovanni, M.T.; de Menezes, J.T.; Bolelli, G.; Cerri, E.; Castrodeza, E.M. Fatigue crack growth behavior of a selective laser melted AlSi10Mg. *Eng. Fract. Mech.* **2019**, *217*, 106564. [CrossRef]

29. Awd, M.; Siddique, S.; Walther, F. Microstructural damage and fracture mechanisms of selective laser melted Al-Si alloys under fatigue loading. *Theor. Appl. Fract. Mech.* **2020**, *106*, 102483. [CrossRef]

30. Sienkiewicz, J.; Płatek, P.; Jiang, F.; Sun, X.; Rusinek, A. Investigations on the mechanical response of gradient lattice structures manufactured via SLM. *Metals* **2020**, *10*, 213. [CrossRef]

31. Xiao, L.; Song, W.; Wang, C.; Liu, H.; Tang, H.; Wang, J. Mechanical behaviour of open-cell rhombic dodecahedron Ti-6Al-4V lattice structure. *Mater. Sci. Eng. A* **2015**, *640*, 375–384. [CrossRef]

32. Maconachie, T.; Leary, M.; Lozanovski, B.; Zhang, X.; Qian, M.; Faruque, O.; Brandt, M. SLM lattice structures: Properties, performance, applications and challenges. *Mater. Des.* **2019**, *183*, 108137. [CrossRef]

33. Boniotti, L.; Beretta, S.; Patriarca, L.; Rigoni, L.; Foletti, S. Experimental and numerical investigation on compressive fatigue strength of lattice structures of AlSi7Mg manufactured by SLM. *Int. J. Fatigue* **2019**, *128*, 105181. [CrossRef]

34. Dallago, M.; Winiarski, B.; Zanini, F.; Carmignato, S.; Benedetti, M. On the effect of geometrical imperfections and defects on the fatigue strength of cellular lattice structures additively manufactured via Selective Laser melting. *Int. J. Fatigue* **2019**, *124*, 348–360. [CrossRef]

35. Zargarian, A.; Esfahanian, M.; Dadkhodapouor, J.; Ziaei Rad, S.; Zamani, D. On the fatigue behavior of additive manufactured lattice structures. *Theor. Appl. Fracture Mech.* **2019**, *100*, 225–232. [CrossRef]

36. Zhang, C.; Zhu, H.; Liao, H.; Cheng, Y.; Hu, Z.; Zeng, X. Effect of heat treatments on fatigue property of selective laser melting AlSi10Mg. *Int. J. Fatigue* **2018**, *116*, 513–522. [CrossRef]

37. Tang, M.; Pistorius, P.C. Oxides, porosity and fatigue performance of AlSi10Mg parts produced by selective laser melting. *Int. J. Fatigue* **2017**, *94*, 192–201. [CrossRef]

38. Muralidharan, U.; Manson, S.S. Modified universal slopes equation for estimation of fatigue characteristics. *Trans. ASME J. Eng. Mater. Tech.* **1988**, *110*, 55–58. [CrossRef]

39. Glodež, S.; Klemenc, J.; Zupanič, F.; Vesenjak, M. The high-cycle fatigue behaviour and fracture analysis of SLM AlSi10Mg alloy. *Trans. Nonferrous Metals Soc. China* **2020**, in press.

40. Ashby, M.F.; Evans, A.; Fleck, N.A.; Gibson, L.J.; Hutchinson, J.W.; Wadley, H.N.G. *Metal Foams: A Design Guide*; Elsevier Science: Burlington, MA, USA, 2000.

41. Lehmhus, D.; Vesenjak, M.; de Schampheleire, S.; Fiedler, T. From stochastic foam to designed structure: Balancing cost and performance of cellular metals. *Materials* **2017**, *10*, 922. [CrossRef]

42. Lim, T.-C. *Auxetic Materials and Structures*; Springer: Berlin/Heidelberg, Germany, 2015; ISBN 978-981-287-274-6.

43. Lakes, R.S. Foam structures with a negative poisson' s ratio. *Science* **1987**, *235*, 1038–1040. [CrossRef]

44. Salit, V.; Weller, T. On the feasibility of introducing auxetic behavior into thin-walled structures. *Acta Mater.* **2009**, *57*, 125–135. [CrossRef]

45. Novak, N.; Vesenjak, M.; Ren, Z. Auxetic cellular materials—A Review. *Strojniški Vestn. J. Mech. Eng.* **2016**, *62*, 485–493. [CrossRef]

46. Yao, Y.T.; Uzun, M.; Patel, I. Workings of auxetic nano-materials. *J. Achiev. Mater. Manuf. Eng.* **2011**, *49*, 585–593.

47. Schwerdtfeger, J.; Heinl, P.; Singer, R.F.; Körner, C. Auxetic cellular structures through selective electron-beam melting. *Phys. Status Solidi* **2010**, *247*, 269–272. [CrossRef]

48. Grima-Cornish, J.N.; Grima, J.N.; Attard, D. A novel mechanical metamaterial exhibiting auxetic behavior and negative compressibility. *Materials* **2019**, *13*, 79. [CrossRef]

49. Schwerdtfeger, J.; Schury, F.; Stingl, M.; Wein, F.; Singer, R.F.; Körner, C. Mechanical characterisation of a periodic auxetic structure produced by SEBM. *Phys. Status Solidi* **2012**, *249*, 1347–1352. [CrossRef]

50. Liu, L.; Hu, H. A review on auxetic structures and polymeric materials. *Sci. Res. Essays* **2010**, *5*, 1052–1063.

51. Duncan, O.; Allen, T.; Foster, L.; Senior, T.; Alderson, A. Fabrication, characterisation and modelling of uniform and gradient auxetic foam sheets. *Acta Mater.* **2017**, *126*, 426–437. [CrossRef]

52. Dobnik Dubrovski, P.; Novak, N.; Borovinšek, M.; Vesenjak, M.; Ren, Z. In-plane behavior of auxetic nonwoven fabrics based on rotating square unit geometry under tensile load. *Polymers* **2019**, *11*, 1040. [CrossRef] [PubMed]

53. Ng, W.S.; Hu, H. Woven fabrics made of auxetic plied yarns. *Polymers* **2018**, *10*, 226. [CrossRef]

54. Zulifqar, A.; Hu, H. Geometrical analysis of bi-stretch auxetic woven fabric based on re-entrant hexagonal geometry. *Text. Res. J.* **2019**. [CrossRef]

55. McDonald, S.A.; Dedreuil-Monet, G.; Yao, Y.T.; Alderson, A.; Withers, P.J. In situ 3D X-ray microtomography study comparing auxetic and non-auxetic polymeric foams under tension. *Phys. Status Solidi* **2011**, *248*, 45–51. [CrossRef]

56. Novak, N.; Vesenjak, M.; Tanaka, S.; Hokamoto, K.; Ren, Z. Compressive behaviour of chiral auxetic cellular structures at different strain rates. *Int. J. Impact Eng.* **2020**, *141*, 103566. [CrossRef]

57. Imbalzano, G.; Tran, P.; Ngo, T.D.; Lee, P.V. Three-dimensional modelling of auxetic sandwich panels for localised impact resistance. *J. Sandw. Struct. Mater.* **2015**. [CrossRef]

58. Nečemer, B.; Kramberger, J.; Vuherer, T.; Glodež, S. Fatigue crack initiation and propagation in re-entrant auxetic cellular structures. *Int. J. Fatigue* **2019**, *126*, 241–247. [CrossRef]

59. Tomažinčič, D.; Vesenjak, M.; Klemenc, J. Prediction of static and low-cycle durability of porous cellular structures with positive and negative Poisson's ratios. *Theor. Appl. Fract. Mech.* **2020**, *106*, 102479. [CrossRef]

60. Novak, N.; Vesenjak, M.; Ren, Z. Computational simulation and optimisation of functionally graded auxetic structures made from inverted tetrapods. *Phys. Status Solidi B* **2017**, *254*. [CrossRef]

61. Al-Rifaie, H.; Sumelka, W. The Development of a new shock absorbing uniaxial graded auxetic damper (UGAD). *Materials* **2019**, *12*, 2573. [CrossRef]

62. Novak, N.; Hokamoto, K.; Vesenjak, M.; Ren, Z. Mechanical behaviour of auxetic cellular structures built from inverted tetrapods at high strain rates. *Int. J. Impact Eng.* **2018**, *122*, 83–90. [CrossRef]

63. Foster, L.; Peketi, P.; Allen, T.; Senior, T.; Duncan, O.; Alderson, A. Application of auxetic foam in sports helmets. *Appl. Sci.* **2018**, *8*, 354. [CrossRef]

64. Airoldi, A.; Bettini, P.; Panichelli, P.; Oktem, M.F.; Sala, G. Chiral topologies for composite morphing structures—Part I: Development of a chiral rib for deformable airfoils. *Phys. Status Solidi B* **2015**, *1445*, 1435–1445. [CrossRef]

65. Airoldi, A.; Bettini, P.; Panichelli, P.; Oktem, M.F.; Sala, G. Chiral topologies for composite morphing structures—Part II: Novel configurations and technological processes. *Phys. Status Solidi B* **2015**, *1454*, 1446–1454. [CrossRef]

66. Ngo, T.D.; Kashani, A.; Imbalzano, G.; Nguyen, K.T.Q.; Hui, D. Additive manufacturing (3D printing): A review of materials, methods, applications and challenges. *Compos. Part. B Eng.* **2018**. [CrossRef]

67. Borovinšek, M.; Novak, N.; Vesenjak, M.; Ren, Z.; Ulbin, M. Designing 2D auxetic structures using multi-objective topology optimisation. *Mater. Sci. Eng. A* **2020**, in press.

68. Ngnekoua, J.N.D.; Nadot, Y.; Henaff, G.; Nicolai, J.; Kan, W.H.; Cairney, J.M.; Ridosz, L. Fatigue properties of AlSi10Mg produced by additive layer manufacturing. *Int. J. Fatigue* **2019**, *119*, 160–172. [CrossRef]

69. Ansys® Academic Research Mechanical Release; ANSYS, Inc. Southpointe, 2600 Ansys Drive: Canonsburg, PA, USA, 2020; R1 (20.1.0.2019111719). Available online: https://www.ansys.com/ (accessed on 26 May 2020).

70. Stephens, R.I.; Fatemi, A.; Stephens, R.R.; Fuchs, H.O. *Metal. Fatigue in Engineering*; John Wiley & Sons Inc.: New York, NY, USA, 2001.

71. MATLAB ver. R2020a (9.8.0.1323502) win64; The MathWorks Inc., 1 Apple Hill Drive: Natick, MA 01760-2098, USA, 2020. Available online: https://www.mathworks.com/ (accessed on 26 May 2020).

Article

Ultrasonic and Conventional Fatigue Endurance of Aeronautical Aluminum Alloy 7075-T6, with Artificial and Induced Pre-Corrosion

Ishvari F. Zuñiga Tello [1], Marijana Milković [2], Gonzalo M. Domínguez Almaraz [1,*] and Nenad Gubeljak [2]

1   Faculty of Mechanical Engineering, University of Michoacan (UMSNH), Santiago Tapia 403, Morelia 58000, Mexico; isfernanda@umich.mx
2   Faculty of Mechanical Engineering, University of Maribor, Smetanova 17, 2000 Maribor, Slovenia; marijana.milkovic@um.si (M.M.); nenad.gubeljak@um.si (N.G.)
*   Correspondence: dalmaraz@umich.mx; Tel.: +52-4433223500-3105

Received: 4 July 2020; Accepted: 24 July 2020; Published: 1 August 2020

Abstract: Ultrasonic and conventional fatigue tests were carried out on the AISI-SAE AA7075-T6 aluminum alloy, in order to evaluate the effect of artificial and induced pre-corrosion. Artificial pre-corrosion was obtained by two hemispherical pitting holes of 500-μm diameter at the specimen neck section, machined following the longitudinal or transverse direction of the testing specimen. Induced pre-corrosion was achieved using the international standard ESA ECSS-Q-ST-70-37C of the European Space Agency. Specimens were tested under ultrasonic fatigue technique at frequency of 20 kHz and under conventional fatigue at frequency of 20 Hz. The two applied load ratios were: R = −1 in ultrasonic fatigue tests and R = 0.1 in conventional fatigue tests. The main results were the effects of artificial and induced pre-corrosion on the fatigue endurance, together with the surface roughness modification after the conventional fatigue tests. Crack initiation and propagation were analyzed and numeric models were constructed to investigate the stress concentration associated with pre-corrosion pits, together with the evaluation of the stress intensity factor in mode I from crack initiation to fracture. Finally, the stress intensity factor range threshold ΔK_{TH} was obtained for the base material and specimens with two hemispherical pits in transverse direction.

Keywords: Al alloy 7075-T6; ultrasonic fatigue; artificial pits; pre-corrosion; crack initiation

1. Introduction

One of the most versatile metal used in industrial applications is aluminum and its alloys; it is the second widely produced metal after steel. Particularly, its alloys are often used in modern industries: more than 40% of aerospace and aeronautical parts are made of these alloys nowadays [1–4]. Frequently, such industrial applications imply a combination of mechanical loading and corrosion, which leads to stress concentration and failure of the material. Pitting corrosion is one of the most damaging effect on materials, difficult to predict in order to avoid damage and failure in industrial components [5–9]. In previous numeric investigations [10], it was found that orientation of the corrosion pits regarding the applied load plays a significant role on the stress concentration factors. An important number of studies has been focused to understanding how cracks initiate and propagate in materials influenced by external factors, such as corrosion pitting, surface conditions, temperature and others [10–13]. A main motivation to carry out this research work is that no studies have been developed concerning the effect of pre-corrosion on the fatigue endurance, under conventional and ultrasonic tests, on this aeronautical aluminum alloy. Concerning surface roughness, fatigue life has been studied in a carbon steel, showing that this superficial parameter influences fatigue life only during crack initiation stage [14]; others investigations [15,16], have referred that fatigue life is reduced with the increase of roughness. In this study, the roughness has been registered before and after the conventional fatigue tests, in order to record its variation.

On the other hand, several studies have pointed out the correlation between pitting corrosion and the reduction of fatigue life [17–20], taking the maximum or the average pit size as the initial crack size; nevertheless, no consideration of pits interactions has been considered inducing high values of stress concentrations factors. In this study, a numeric model reproducing two pits on the specimen surface is developed in order to investigate the increase of stress concentrations factors associated with its orientation. Fatigue endurance of pitting metals can be improved applying shoot peening, which induces compression residual stress [21,22].

2. Materials and Methods

The chemical composition (% in weight) and the main mechanical properties of AA7075-T6 aluminum alloy are listed in Tables 1 and 2, respectively. In order to obtain experimentally some mechanical properties as yield stress and ultimate tensile stress of this aluminum alloy (Table 2), tests were conducted on universal testing machine Instron 1255 (Instron, Norwood, NY, USA) on 4 specimens (cold-rolled Al7075-T6 sheet, tension applied parallel to the rolling direction of the sheet) of non-standard dimensions and at room temperature.

Table 1. Chemical composition by weight (%) of aluminum alloy 7075-T6.

Zn.	Mg	Cu	Cr	Fe	Al
6.9 Max.	2.7 Max.	1.87 Max.	0.2 Max.	0.4 Max.	Balance

Table 2. Main mechanical properties of aluminum alloy 7075-T6.

Density	Hardness	σ_y	σ_u	Poisson Ratio	Elastic Modulus
(kg/m^3)	(HV0.1)	(MPa)	(MPa)	(–)	(GPa)
2800	155.12	453	538	0.33	70

The specimens used in ultrasonic fatigue test were machined from an AA7075-T6 plate, its dimensions were determined by modal numeric analysis in order to fit the resonance condition: the natural frequency of vibration in longitudinal direction close to frequency of excitation source (20 kHz ± 300 Hz). The dimensions (mm) are shown in Figure 1a for the ultrasonic specimen and, Figure 1b for the conventional specimen. In Figure 2 is depicted the result of modal analysis: 19,929 Hz of natural frequency for ultrasonic specimen with dimensions presented in Figure 1a, using the mechanical properties of Table 2: density, Young's modulus and Poisson ratio. Ansys 19-R1 was used with 12,120 elements, 55,753 nodes and a fixed support at the specimen's end as boundary condition.

Specimens used for conventional fatigue tests present similar dimensions of the previous ultrasonic specimens, slightly modifications were introduced, such as: a machined rope at the extremes of specimen in order to fix it to the testing machine, Figure 1b.

Experimental Set Up

All ultrasonic fatigue tests were performed with the self-designed and constructed machine Figure 3, which is totally controlled by a LabVIEW program (Version 2015, National Instruments, Austin, TX, USA), allowing the test initiation, the record of the number of cycles in real time and the automatic stop with the specimen's failure. Calibration of displacements of specimens at the free end was carried out by an inductive proximity sensor (Keyence, Itasca, IL, USA), which has a resolution of ±2 µm, working at 1.5 MHz. All tests were obtained at room temperature (close to 23 °C), with environmental humidity between 35% to 55% and full reversed load ratio R = −1.

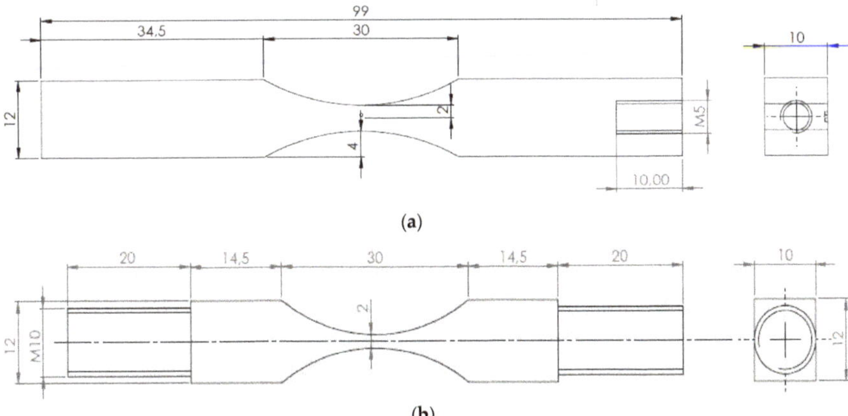

Figure 1. (a) Ultrasonic fatigue specimen dimensions (mm), fitting the resonance condition; (b) dimensions (mm), of the conventional fatigue specimen.

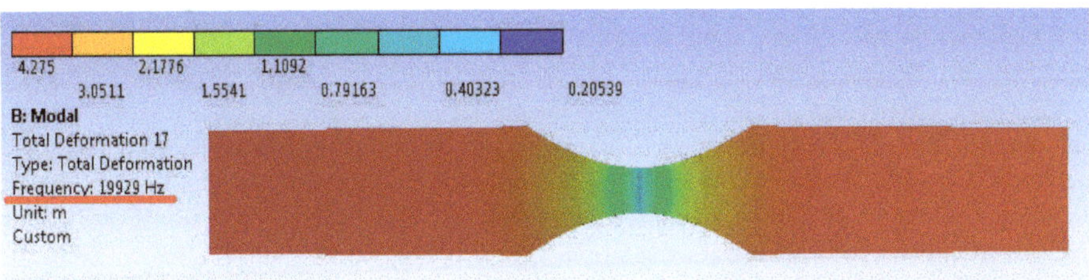

Figure 2. Natural frequency of vibration in longitudinal direction for aluminum alloy specimen AA7075-T6, obtained by modal analysis.

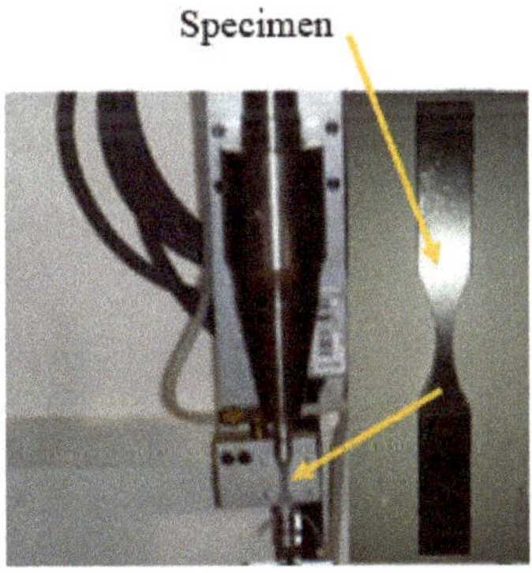

Figure 3. Ultrasonic testing machine.

The experimental start up for each specimen under ultrasonic fatigue tests was as follows: stabilization of the system for 30 s with an applied load of 44 MPa (corresponding to the lower voltage of ultrasonic generator: 10 volts); afterwards, increasing 4.4 MPa each second to attain the desired applied load.

Conventional fatigue tests were conducted on a servo–hydraulic testing machine, in which frequency can be imposed from 0.5 to 20 Hz. The complete system of the machine was developed in the University of Maribor, Slovenia [23]; an overview is depicted in Figure 4. The manufacturing of the parts and the whole controlling system for the machine was configured to work between 0.5 to 20 Hz; all tests under conventional fatigue were obtained at the highest frequency and with load ratio R = 0.1, at room temperature without control of environmental humidity.

Figure 4. Servo-hydraulic machine for conventional fatigue testing.

Roughness on surface of testing specimens was measured before and after the conventional fatigue tests, in order investigate the variation of this parameter. The measurements were obtained in longitudinal and transversal direction in regards the principal axis of specimens, using two roughness measuring devices: a TESA Rugosurf 10 G (Renens, Switzerland) and a Mitutoyo SURFTEST SJ-210 (Queréraro, Mexico). The use of these two devices allows to corroborate similarity of the average roughness obtained through the first and second MEASURING device.

Three initial types of specimens were tested: specimens with nomination A, which are the base material, nomination B with two artificial hemispherical pits (500-μm diameter) in transverse direction in regard the applied load, Figure 5 and nomination C with two artificial hemispherical pits (500-μm diameter), this time longitudinally oriented in regard the applied load, Figure 6. Specimens to be machined at the neck section are presented in Figure 7a (milling machine, model Dyna EM-3116, San José, CA, USA); whereas the two artificial pits were made with a tungsten drill bit of 500-μm-diameter (brand: HANGXIN, model: ZT0.5, Guangzhou, China), as illustrated in Figure 7b. Pre-corrosion was achieved by immersion of specimens during 30 h in a solution of distilled water with 3.5% in weight of NaCl (international standard ESA ECSS-Q-ST-70-37C). Figure 8a presents the process of immersion for pre-corrosion and Figure 8b, the microstructure of pre-corroded specimen on the flat surface, with a higher pit dimension of about 90 μm, approximately.

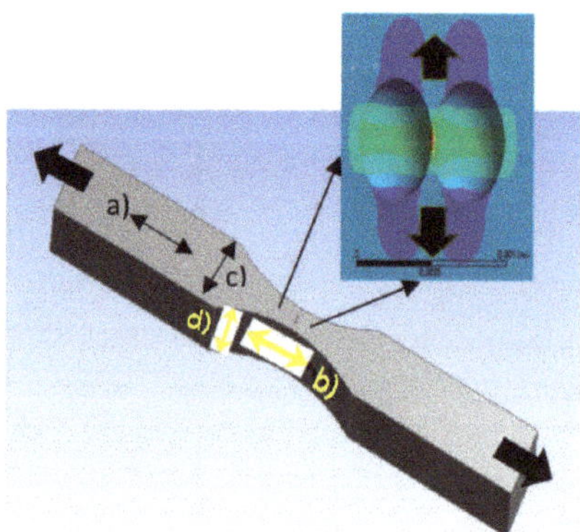

Figure 5. Specimen B—transverse pits.

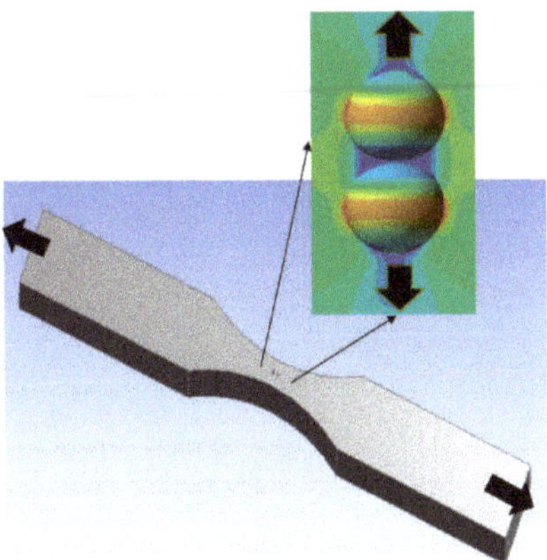

Figure 6. Specimen C—longitudinal pits.

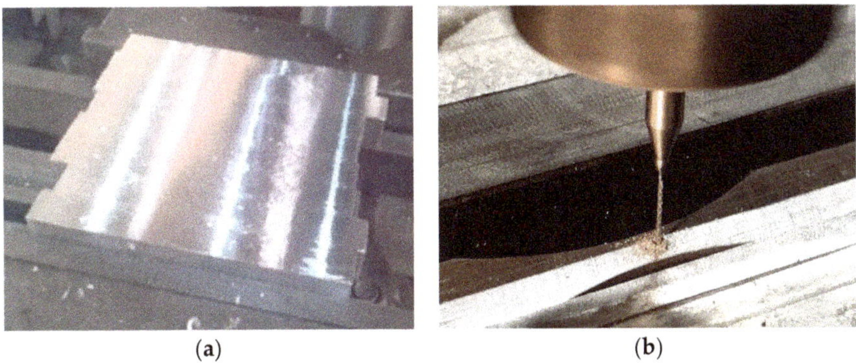

(a)   (b)

Figure 7. (**a**) Specimens to be machined at the neck section; (**b**) drilling of the two artificial pits holes.

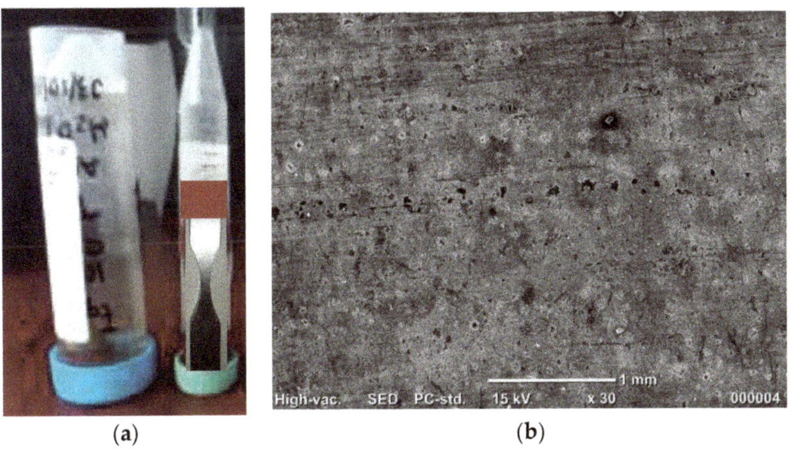

(a)   (b)

Figure 8. (**a**) Pre-corrosion process; (**b**) pre-corroded surface at the flat side of specimen.

3. Results

3.1. Fatigue Endurance under Ultrasonic and Conventional Techniques

Ultrasonic and conventional fatigue tests on the base material, with two pits and pre-corroded specimens are plotted on Figure 9 (the maximum Von-Mises stresses of fractured specimens).

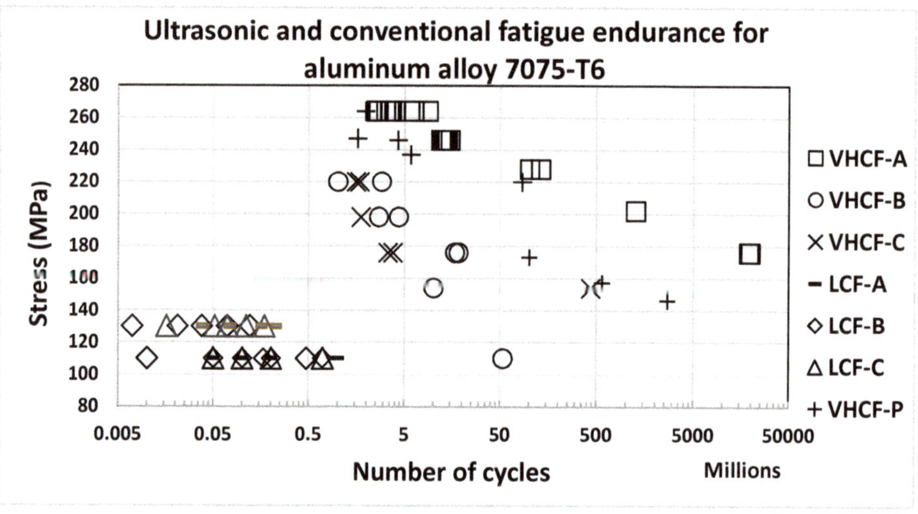

Figure 9. Results of ultrasonic and conventional fatigue tests.

Nomination for ultrasonic fatigue were as follows: VHCF-A the base material, VHCF-B for specimens with two artificial pitting holes in transverse direction, VHCF-C specimens with two artificial pitting holes in longitudinal direction and VHCF-P for specimens with pre-corrosion. Nomination in conventional fatigue was: LCF-A the base material specimens, LCF-B specimens with two artificial pitting holes in transverse direction and LCF-C specimens with two artificial pitting holes in longitudinal direction.

3.2. Roughness Measurements on the Conventional Fatigue Testing Specimens

Measurements of surface roughness were obtained before and after the conventional fatigue tests, as shown in Table 3.

The roughness parameter Ra was measured in four localizations: (a) longitudinal direction at the flat surface, (b) longitudinal direction at the side of specimen, (c) transverse direction at the flat surface and (d) transverse direction at the side of specimen, as illustrated in Figure 5. The obtained results show that roughness slightly decreases in all the specimens with nomination A (without holes) and in the specimens with nomination B and C (transverse and longitudinal pits, respectively), in the direction of the applying load and slightly increases in the transverse direction. The last behavior may be related to microplastic deformation in tension for the longitudinal direction and compression for the transverse direction.

In Table 3 are listed measured values of roughness before * and after ** the conventional fatigue tests; together with testing parameters: NT the target number of cycles, NF the real number of cycles, F_{max} the value of maximum force, σ_{max} the maximum stress in the specimen, R the load ratio and f the frequency of the cyclic loading.

Table 3. Roughness on specimens of aluminum alloy 7075-T6, before * and after ** the conventional fatigue tests.

#	N_T	N_F	F_{max} (KN)	σ_{max} (MPa)	R	f (Hz)	Ra Localization (a) (µm)	Ra Localization (b) (µm)	Ra Localization (c) (µm)	Ra Localization (d) (µm)
A1	10^7	382,956	4.86	110	0.1	20	* 0.179	0.252	0.225	0.705
							** 0.144	0.358	0.292	0.230
A2	50,000	50,000	4.75	110	0.1	20	* 0.141	0.409	0.322	0.643
A3	10^7	460,002	4.39	110	0.1	20	* 0.127	0.381	0.223	0.578
							** 0.125	0.244	0.220	0.558
A4	10^7	354,215	3.49	110	0.1	20	* 0.17	0.323	0.193	0.871
							** 0.176	0.298	0.224	0.548
A5	10^7	10^7	2.31	110	0.1	20	* 0.103	0.556	0.197	0.747
							** 0.09	0.283	0.182	0.660
A6	10^5	10^5	2.25	110	0.1	20	* 0.105	0.48	0.145	0.253
							** 0.113	0.441	0.158	0.606
A7	200,000	200,000	2.34	110	0.1	20	* 0.128	0.379	0.138	0.590
							** 0.104	0.258	0.185	0.575
A8	10^7	10^7	2.27	110	0.10	20	* 0.11	0.628	0.164	0.619
							** 0.103	0.331	0.158	0.590
B1	10^6	692,023	4.24	110	0.1	20	* 0.31	0.538	0.4	0.731
							** 0.271	0.375	0.755	0.999
B2	10^4	10^4	4.88	110	0.1	20	* 0.794	0.663	0.547	0.487
B3	10^7	10^7	2.29	110	0.10	20	* 0.166	0.518	0.386	0.830
							** 0.113	0.423	0.467	0.963
B4	10^7	165,047	3.38	110	0.1	20	* 0.228	0.467	0.403	0.866
							** 0.145	0.393	0.342	0.885
B5	50,000	50,000	2.31	110	0.1	20	* 0.128	0.409	0.25	0.55
							** 0.103	0.447	0.219	0.623
B6	10^5	10^5	2.29	110	0.10	20	* 0.119	0.599	0.18	0.633
							** 0.125	0.450	0.189	0.654
B7	200,000	200,000	2.33	110	0.1	20	* 0.114	0.434	0.371	0.776
							** 0.106	0.275	0.214	0.838
C1	10^7	335,033	4.58	110	0.1	20	* 0.244	0.586	0.382	0.98
							** 0.159	0.596	0.395	1.091
C3	10^7	10^7	2.28	110	0.10	20	* 0.163	0.483	0.336	0.912
							** 0.149	0.288	0.278	0.530
C4	700,000	700,000	2.29	110	0.1	20	* 0.166	0.643	0.265	0.812
							** 0.17	0.486	0.273	0.909
C5	50,000	50,000	2.22	110	0.1	20	* 0.168	0.580	0.383	0.877
							** 0.127	0.450	0.339	0.672
C6	10^5	10^5	2.29	110	0.10	20	* 0.155	0.538	0.279	0.709
							** 0.19	0.461	0.265	0.657
C7	200,000	200,000	2.32	110	0.10	20	* 0.167	0.582	0.248	0.712
							** 0.14	0.286	0.282	0.715

3.3. Microstructure of Crack Initiation and Propagation in Pre-Corroded Specimens and with Two Artificial Pits

Crack initiation for the pre-corroded specimens was frequently localized in one or some close pre-corrosion pits, as shown in Figure 10a; whereas crack propagation was systematically developed at the neck section of specimens, following an intergranular path, as illustrated in Figure 10b.

Figure 10. (**a**) Crack initiation in a pre-corrosion pit; (**b**) crack propagation following intergranular path.

For the two artificial pitting holes, the crack initiates at the bottom of pits in the case of transverse pits, Figure 11a and it is localized at the bottom of one pit in the case of longitudinal pits, Figure 11b. A numeric analysis is developed in the next section in order to investigate the stress concentration factors induced by the two artificial pits in the transverse or longitudinal direction and its impact in fatigue endurance.

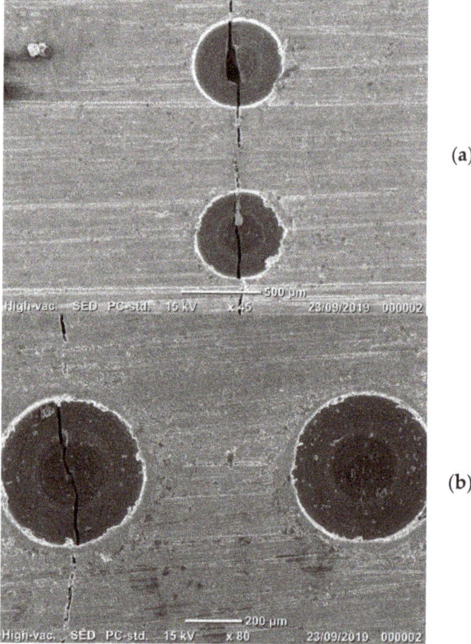

Figure 11. (**a**) Crack initiation and propagation for two transverse pits; (**b**) crack initiation and propagation for two longitudinal pits.

3.4. Numeric Simulation of Stress Concentration

In order to assess the effect of artificial pits (two hemispherical pits oriented in transverse and longitudinal direction in regard the applied load), numeric simulations were carried out on the testing specimen. In Figure 12 is illustrated the stress distribution on the testing specimen without pits, undergoing a total displacement of 130 µm. The maximum Von Mises stress is located at the neck section with the value of 220 MPa (corresponding to the maximum applied stress in Figure 9 for VHCF-B and VHCF-C).

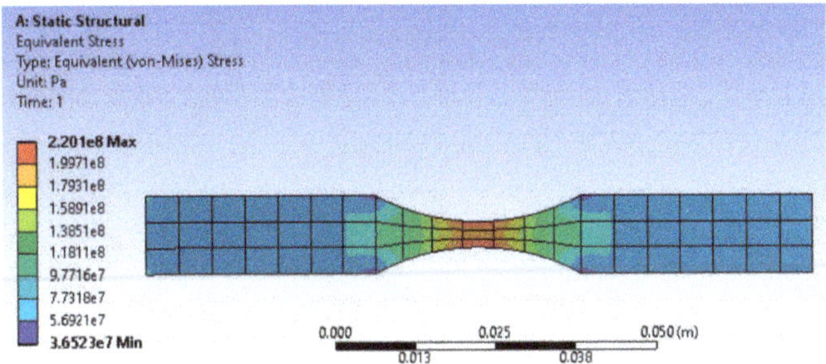

Figure 12. Stress distribution of testing specimen with 130 µm of displacement at the end.

The numeric results for the specimen with two hemispherical pits in transverse direction and separated 500 µm, are presented in Figure 13a. Physical fracture in this type of specimen is shown in Figure 13b; the maximum numeric Von-Mises stress was 358 MPa, approximately.

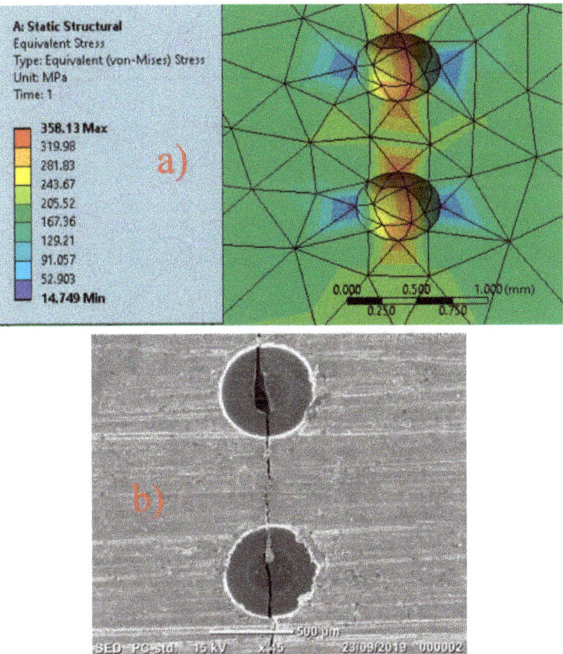

Figure 13. (a) Stress distribution for two pits in transverse direction and 130 µm of total displacement; (b) fracture on two pits in transverse direction.

The stress distribution for the specimen with two pits separated 500 µm, this time in longitudinal direction in regard the applied load, is illustrated in Figure 14a. The maximum numeric Von-Mises stress was 320 MPa. Comparing Figures 12–14, the stress concentration factor is: 358/220 = 1.63 for two pits in transverse direction and 320/220 = 1.45 for the two pits in longitudinal direction. Stress concentration factors was obtained by numeric simulation in different materials and conditions [24–26]. Numeric analysis in Figures 12–14 was obtained with the Ansys software (Version R19, Ansys Inc., Canonsburg, PA, USA), SOLID187 element type and 36,039 and 35,945 elements, respectively. One end side of specimen was fixed as boundary condition.

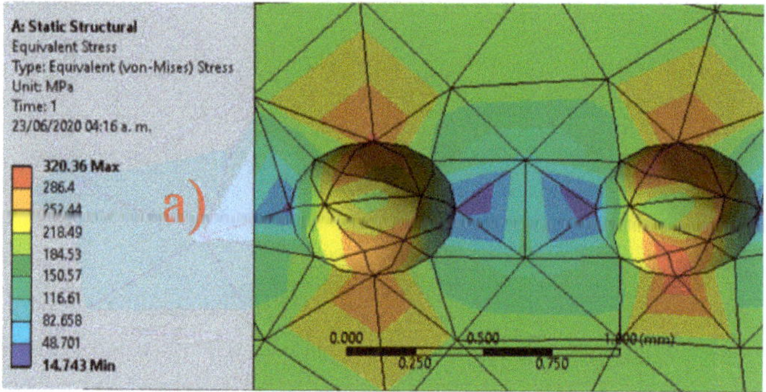

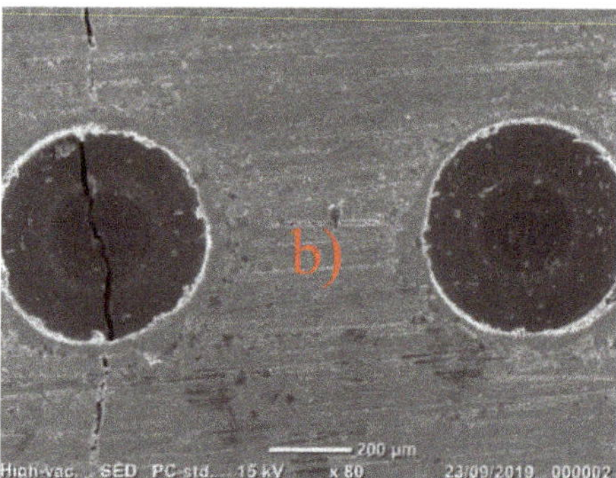

Figure 14. (**a**) Stress distribution for two pits in longitudinal direction and 130 µm of total displacement; (**b**) fracture on two pits in longitudinal direction.

3.5. Numeric Determination of Stress Intensity Factor in Mode I

The stress intensity factor (SIF) in mode I during the crack propagation of aluminum alloy 7075-T6 was evaluated assuming as crack initiation a hemispherical pit generated at the surface, Figure 15.

The 3D numeric simulation can be reduced to a 2D numeric simulation under the following considerations: the initial crack length is small (400 µm), compared with the specimen dimensions so that plane stress is assumed. The 2D model is illustrated in Figure 16a, with the initial crack length localized at the bottom of one pit, as shown in Figure 16b. Quadrilateral 2D elements was used: 25,636 elements and 26,190 nodes; fixed one vertical side of specimen for boundary conditions.

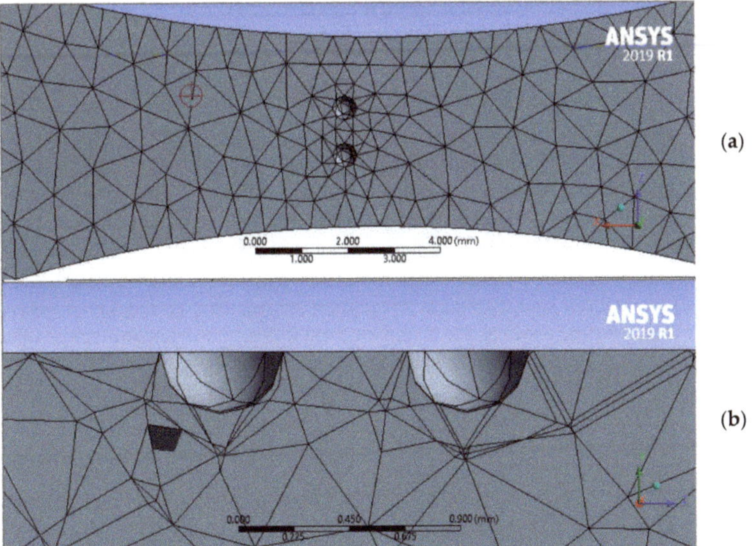

Figure 15. (**a**) Numeric 3D model for evaluate the stress intensity factor in mode I, (**b**) section view.

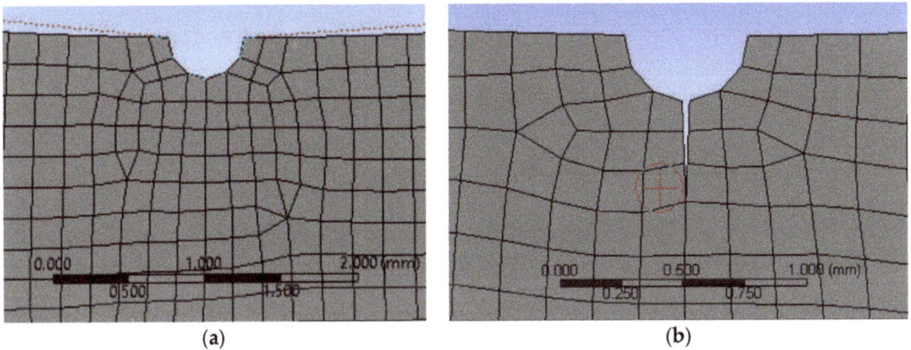

Figure 16. (**a**) 2D numerical model to evaluate SIF in mode I; (**b**) initial crack length of 400 μm.

Crack growth is determined as straight-line segments across the 2D body in the interval 0–S, divided into steps Δsi, $i = 1 \ldots$ n. For a determined state of stress and displacement $(\sigma, u)^i$, at the step i for the crack geometry E^i, the crack growth evaluation consists to determine the geometry E^{i+1}, corresponding to $i+1$ step.

The propagation angle ϕ^{i+1} and the incremental length Δa^{i+1} for each step must be obtained; the selected criteria for the crack growth direction was the normal to the direction of maximum hoop stress [27,28]. Assuming 2D loading and cylindrical coordinate system at the crack tip, the asymptotic circumferential and shear stresses in this region are represented by:

$$\sigma_{\varphi\varphi} = \frac{K_I}{\sqrt{2\pi r}} \frac{1}{4} \left(3\cos\left(\frac{\varphi}{2}\right) + \cos\left(\frac{3\varphi}{2}\right)\right) + \frac{K_{II}}{\sqrt{2\pi r}} \frac{1}{4} \left(-3\sin\left(\frac{\varphi}{2}\right) - 3\sin\left(\frac{3\varphi}{2}\right)\right) \tag{1}$$

$$\sigma_{r\varphi} = \frac{K_I}{\sqrt{2\pi r}} \frac{1}{4} \left(\sin\left(\frac{\varphi}{2}\right) + \sin\left(\frac{3\varphi}{2}\right)\right) + \frac{K_{II}}{\sqrt{2\pi r}} \frac{1}{4} \left(\cos\left(\frac{\varphi}{2}\right) + 3\cos\left(\frac{3\varphi}{2}\right)\right) \tag{2}$$

Equation (1) represents the principal stress in the direction of crack propagation. The shear stress $\sigma_{r\theta}$ is assumed vanishing in this direction; then, it is obtained:

$$\sigma_{r\varphi} = 0 = \frac{1}{\sqrt{2\pi r}}\ \cos\left(\frac{\varphi}{2}\right)\left[\frac{K_I}{2}\left(\sin\left(\varphi\right) + \frac{K_{II}}{2}\left(3\cos\varphi - 1\right)\right)\right] \tag{3}$$

Here: ϕ is the angle of crack propagation in the tip coordinate system.

The mixed mode can be approached to single mode loading, which leads to: $K_{II} \approx 0$ and equation 3 becomes: $K_I \sin(\phi) = 0$. The last result implies that $\phi = 0$ for each step along the crack propagation. Experimental observations seem confirm this result: crack initiation and propagation are developed at the neck section of specimens, perpendicular to the longitudinal axis of specimen, as observed in Figure 11a,b. The numeric domain was constructed with 25,636 quadrilateral elements and 26,190 nodes. K_I was evaluated using the displacement correlation method [29,30], with the displacement vectors in the X direction Vi and the equation:

$$K_I = \frac{G}{k+1}\sqrt{\frac{2\pi}{L}}\left[4(V_1 - V_2) - (V_3 - V_4)\right] \tag{4}$$

Figure 17 illustrates the parameters involved for the numeric evaluation of K_I. The mechanical properties were shear modulus G = 26 GPa, k = (3–4v) for plane stress, L is the length of the singular element and v = 0.33 the Poisson's coefficient.

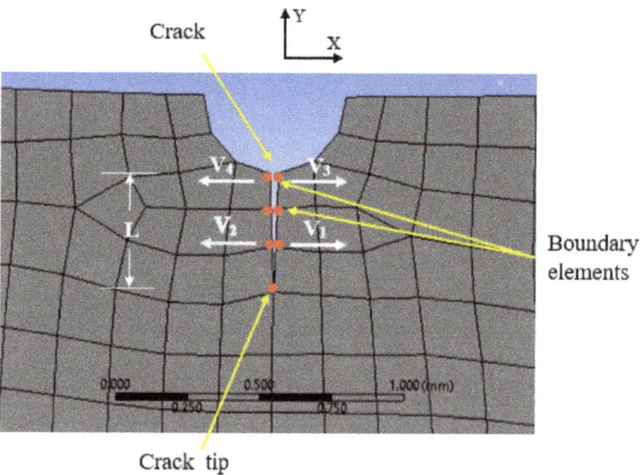

Figure 17. Crack propagation parameters used in numeric simulation.

The numeric stress intensity factor in mode I was obtained taking the following sequence:

- Determination of stress and strain distributions on cracked structure;
- Evaluation of displacements vectors at the crack zone on the X direction, to calculate the stress intensity factor in mode I;
- Compute the direction (φ = 0 for all steps), of crack extension and the incremental crack surface construction;
- Compute the number of load cycles of last increment for crack growth simulation;
- Re-meshing of boundary, if necessary;
- Return to first point.

K_I was computed from a_0 = 400 μm to a_{end} = 4000 μm (the wide at the specimen neck section). Figure 18 shows the evolution of K_I determined by FEA, using this method.

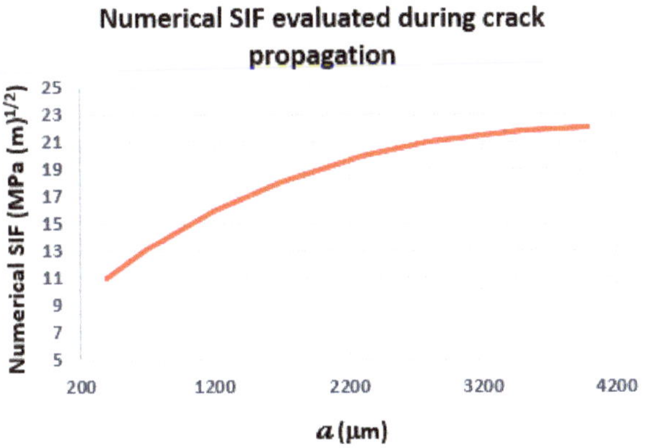

Numerical SIF evaluated during crack propagation

Figure 18. Numeric stress intensity factor (SIF) evaluation in mode I by the displacement correlation method.

3.6. Evaluation of ΔK_{TH}, for Base Material and Specimens with Two Hemispherical Pits in Transverse Direction

The stress intensity factor range threshold ΔK_{TH} was investigated for the base material and specimens with two hemispherical pits in transverse direction regarding its longitudinal axis.

The initial crack size is small compared with the specimen dimension; then, the stress intensity factor in mode I can be obtained with the approximation of edge crack in a semi-infinite body, with the expression:

$$\Delta K_I = Y \sqrt{\pi a} \, (\Delta \sigma) \tag{5}$$

With: Y the geometry correction factor, $Y = 1.12$. Ultrasonic fatigue tests were carried out with $R = -1$; the compressive stress does not make contribution toward the advancement of the fatigue crack. Under this condition, the stress intensity factor range was computed from $\Delta \sigma / 2$. Ultrasonic fatigue testing in the base material specimens presented no crack propagation using the stress amplitude $\Delta \sigma / 2 = 95$ MPa, with an initial crack length $a_0 = 400$ μm; then, Equation (5) leads to: $\Delta K_{TH} \approx 3.77$ MPa (m)$^{0.5}$.

The crack growth behavior can be represented in the stable crack propagation zone by the Paris-Gomez-Anderson law:

$$\frac{da}{dN} = C(\Delta K)^n \tag{6}$$

The constant C and n are properties of tested material, determined by experimental results. For the base material of 7075-T6 alloy at room temperature and $R = -1$, the corresponding values were: $C = 0.5 \times 10^{-12}$ and $n = 5$ and for specimens with two hemispherical pits in transverse direction, these values were: $C = 3.3 \times 10^{-12}$, $n = 5.8$. In Figure 19 are plotted the points da/dN-ΔK for the two specimens in the log scale.

Values obtained for ΔK_{TH} were: ≈ 4 MPa (m)$^{0.5}$ and ≈ 1.8 MPa (m)$^{0.5}$ for the base material and with two pits, respectively. Reported values for ΔK_{TH} with $R = -1$ and room temperature yielded close to 4 MPa (m)$^{0.5}$ [31]. This alloy coated with cladding layer under similar conditions as before and using the effective stress-intensity factor range (ΔK_{eff}), leads to a threshold comprised between 1 and 4 MPa (m)$^{0.5}$ [32]. The value of ΔK_{TH} for this aluminum alloy was reported decreasing when the load ratio R increases, from $R = -1$, $R = 0$ and $R = 0.5$, regardless of taking ΔK or ΔK_{eff} [33].

On the other hand, the number of cycles in the stable zone of crack growth, may be estimate by the Paris–Gomez–Anderson equation:

$$N_{II} = \int_{a_0}^{a_{end}} \frac{da}{C(\Delta K)^n} = \int_{a_0}^{a_{end}} \frac{da}{C\left[Y(a)\Delta \sigma \sqrt{\pi a}\right]^n} \tag{7}$$

With: N_{II} the number of cycles in this zone, $Y(a)$ the geometry factor, taken constant along the crack propagation $Y(a) = 1.12$, a_0 and a_{end} the initial and the final crack length, respectively. Assuming that: $\Delta K = \Delta K_0 (a/a_0)^{1/2}$, together with: $\Delta K_0 = Y(a_0)\, \Delta\sigma\, (\pi a_0)^{1/2}$, $a_0 = 400$ µm, $a_{end} = 4000$ µm, Equation (7) becomes:

$$N_{II} = \frac{2a_0}{(n-2)C[\Delta K_0]^n} \left[1 - \left[\frac{a_0}{a_{end}} \right]^{\frac{n}{2}-1} \right] \tag{8}$$

With: $\Delta K_0 \approx \Delta K_{TH} \approx 3.77$ MPa $(m)^{0.5}$, $n = 5$ and $C = 0.5 \times 10^{-12}$, the number of cycles of base specimen in this zone is: 678,162 cycles, less than 7% of the total fatigue life ($\approx$10 million of cycles at 245 MPa, Figure 9), For the second case and using the values: $C = 3.3 \times 10^{-12}$ and $n = 5.8$, the number of cycles in the stable zone of crack growth is reduced to 28,609 cycles, less than 1% of total fatigue life of these specimens (2–3 million of cycles, Figure 9) at 220 MPa [34,35].

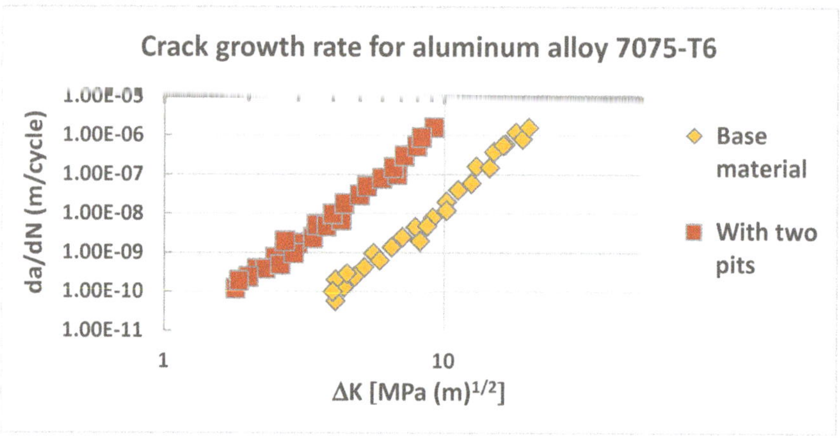

Figure 19. Crack growth rate for the base material and specimens with two pits, under ultrasonic fatigue tests.

4. Discussion

Conventional and ultrasonic fatigue tests were performed on the aluminum alloy 7075-T6 for the base material, specimens with two hemispherical pits disposed in transverse and longitudinal direction in regard the applied load and pre-corroded specimens. Two pits in transverse direction seem induce higher detrimental effect in fatigue endurance associate with the high stress concentration factors, which were determined by numeric simulation. Fatigue endurance for conventional tests revealed similar behavior: the lowest values was observed for specimens with two pits in transverse direction. Ultrasonic fatigue tests on aluminum alloy 7075-T6 were performed under similar condition compared to this study for the nomination VHCF-A [36]. The fatigue endurance was lower compared to the present study for the high load (260 MPa) and similar for the low load (200 MPa); this discrepancy is probably related to difference on the mechanical properties. Concerning measurement of roughness before and after testing under conventional fatigue tests, roughness decreases slightly after fatigue tests in parallel direction to applied load and increases slightly in perpendicular direction to applied load. These results may be associate with tension microplasticity in the first case and compression microplasticity in the second case.

Crack initiates in one or more close pits in the case of pre-corrosion; this localization is at the bottom of one or the two pits, in the case of two artificial pits. Crack initiation is frequently associated with the localization of high stress concentration; physical observations were supported by numeric results. The grain size mean value for this aluminum alloy was 80 µm and crack initiation may be triggered from a single pit occupying a grain or comprising different grains, depending on the crack initiation pit size and localization. The crack propagation in all fatigue tests follows an intergranular path across the specimen neck section; high temperature recorded for conventional fatigue was about 40 °C and close to 31 °C in the case of ultrasonic fatigue, as illustrated in Figure 20, using a thermographic camera.

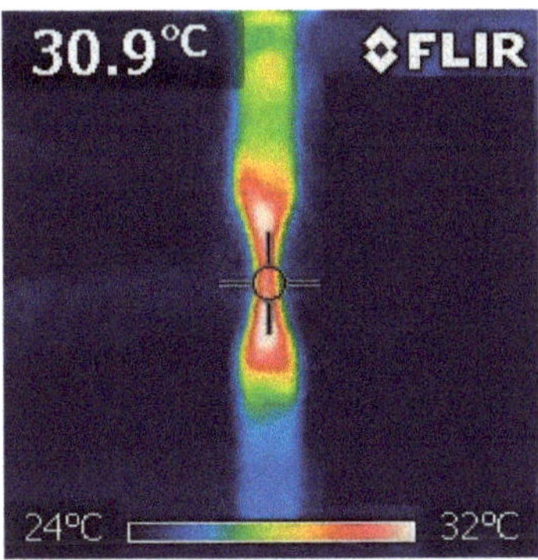

Figure 20. Temperature recorded by thermographic camera, during ultrasonic fatigue test.

The stress intensity factor in mode I was evaluated numerically using the Displacement Correlation Method and the simplification of 2D domain; its values range from: ≈ 11 MPa $(m)^{0.5}$ at crack initiation ($a_0 = 400$ µm) and ≈ 22 MPa $(m)^{0.5}$ at fracture ($a_{end} = 4000$ µm). The stress intensity factor range threshold ΔK_{TH} was obtained for the base material and specimens with two hemispherical pits in transverse direction: ≈ 4 and ≈ 1.8 MPa $(m)^{0.5}$, respectively; these values are of the order found in literature. The reduction of ΔK_{TH} in the second case is related to stress concentrations induced by the two pits. The number of cycles inside the stable zone of crack growth for the base material and specimens with two pits in transverse direction were evaluated, leading to less than 7% of the total ultrasonic fatigue life for the base material and less than 1% for specimens with two pits in transverse direction.

5. Conclusions

The following conclusions can be drawn from the present investigation;

- Ultrasonic fatigue tests were obtained on specimens of the aluminum alloy 7075-T6: base material, with two hemispherical pits and pre-corroded. In the case of conventional fatigue: base material and with two hemispherical pits;
- Pit orientation regarding the applied load plays an important role on the fatigue life of this aluminum alloy: fracture occurs first, for specimen with two transversal pits, following by specimens with longitudinal pits and pre-corroded specimens. The higher fatigue endurance corresponds to specimens of base material;
- Under conventional fatigue tests Ra roughness parameter slightly decreases in the direction of applied load after tests and slightly increases in perpendicular direction;
- Cracks initiate frequently associated with one or more pits for pre-corroded specimens and at the bottom of one or the two artificial pits. Crack propagation was always located across the specimen neck section, following a intergranular path;
- The stress concentration factor Kt, plays a principal role affecting fatigue endurance in the pre-corroded and two pits specimens; values of Kt were obtained by numeric simulation;
- Stress intensity factor in mode I was obtained by numeric evaluation using the displacement correlation method, from crack initiation to fracture;
- The stress intensity factor range threshold was obtained for the base material and for specimens with two hemispherical pits in transverse direction;

- Specimens with pre-corrosion and artificial pits present decrease of the number of fatigue cycles, inside the stable zone of crack growth;
- Additional physical and numeric investigations should be undertaken to assess the quantitative effects of pits in the corroded surfaces—particularly the interaction effects of close pits associated with the crack initiation and propagation.

Author Contributions: I.F.Z.T.: Conception or design of the work, Data collection, Data analysis and interpretation; M.M.: Conception or design of the work, Data collection, Data analysis and interpretation; G.M.D.A.: Conception or design of the work, Data collection, Data analysis and interpretation, Drafting the article, Critical revision of the article; N.G.: Conception or design of the work, Data collection, Data analysis and interpretation, Drafting the article. All authors have read and agreed to the published version of the manuscript.

Funding: This research was funded by Grant number: CB-241117-2014.

Acknowledgments: The authors express their mention of gratitude to CONACYT (The National Council for Science and Technology, Mexico). An additional mention of gratitude to the University of Michoacán in Mexico and the University of Maribor in Slovenia, for the received support in the development of this work.

Conflicts of Interest: The authors declare no conflict of interests.

References

1. Wanhill, R.J.H. Aerospace applications of aluminum-lithium alloys. In *Aluminum-Lithium Alloys, Processing, Properties and Applications*; Prasad, N.E., Gokhale, A.A., Wanhill, R.J.H., Eds.; Butterworth-Heinemann, Elsevier Inc.: Oxford, UK, 2013; pp. 503–535.
2. Zhang, X.; Chen, Y.; Hu, J. Recent advances in the development of aerospace materials. *Prog. Aerosp. Sci.* **2018**, *97*, 22–34. [CrossRef]
3. Jawalkar, C.S.; Kant, S.; Kaushik, Y. A Review on use of Aluminium Alloys in Aircraft Components. *i-Manager's J. Mater. Sci.* **2015**, *3*, 33–38.
4. Rambabu, P.; Eswara Prasad, N.; Kutumbarao, V.V.; Wanhill, R.J.H. Aluminium Alloys for Aerospace Applications. In *Aerospace Materials and Material Technologies*; Indian Institute of Metals Series; Springer: Singapore, 2016; pp. 29–52.
5. Codaro, E.N.; Nakazato, R.Z.; Horovistiz, A.L.; Ribeiro, L.M.F.; Ribeiro, R.B.; Hein, L.D.O. An image analysis study of pit formation on TiÁ 6AlÁ 4V. *Mater. Sci. Eng.* **2002**, *341*, 202–210. [CrossRef]
6. DuQuesnay, D.L.; Underhill, P.R.; Britt, H.J. Fatigue crack growth from corrosion damage in 7075-T6511 aluminium alloy under aircraft loading. *Int. J. Fatigue* **2003**, *25*, 371–377. [CrossRef]
7. Jones, K.; Hoeppner, D.W. The interaction between pitting corrosion, grain boundaries, and constituent particles during corrosion fatigue of 7075-T6 aluminum alloy. *Int. J. Fatigue* **2009**, *31*, 686–692. [CrossRef]
8. Molent, L. Managing airframe fatigue from corrosion pits—A proposal. *Eng. Fract. Mech.* **2015**, *137*, 12–25. [CrossRef]
9. Wang, Q.Y.; Kawagoishi, N.; Chen, Q. Effect of pitting corrosion on very high cycle fatigue behavior. *Scr. Mater.* **2003**, *49*, 711–716. [CrossRef]
10. Zuñiga Tello, I.F.; Domínguez Almaraz, G.M.; López Garza, V.; Guzmán Tapia, M. Numerical investigation of the stress concentration on 7075-T651 aluminum alloy with one or two hemispherical pits under uniaxial or biaxial loading. *Adv. Eng. Softw.* **2019**, *131*, 23–25. [CrossRef]
11. Pao, P.S.; Gill, S.J.; Feng, C.R. On fatigue crack initiation from corrosion pits in 7075-T7351 aluminum alloy. *Scr. Mater.* **2000**, *43*, 391–396. [CrossRef]
12. Stanzl-tschegg, S. Fatigue crack growth and thresholds at ultrasonic frequencies. *Int. J. Fatigue* **2006**, *28*, 1456–1464. [CrossRef]
13. Domínguez Almaraz, G.M.; Ambriz, J.L.Á.; Cadenas Calderón, E. Fatigue endurance and crack propagation under rotating bending fatigue tests on aluminum alloy AISI 6063-T5 with controlled corrosion attack. *Eng. Fract. Mech.* **2012**, *93*, 119–131. [CrossRef]
14. Deng, G.; Nagamoto, K.; Nakano, Y.; Nakanishi, T. *Evaluation of the Effect of Surface Roughness on Crack Initiation Life*; ICF12; Natural Resources Canada: Ottawa, ON, Canada, 2009; pp. 1–8.
15. Singh, K.; Sadeghi, F.; Correns, M.; Blass, T. A microstructure based approach to model effects of surface roughness on tensile fatigue. *Int. J. Fatigue* **2019**, *129*, 105229. [CrossRef]

16. Lai, J.; Huang, H.; Buising, W. Effects of microstructure and surface roughness on the fatigue strength of high-strength steels. *Proc. Struct. Integr.* **2016**, *2*, 1213–1220. [CrossRef]

17. Sankaran, K.K.; Perez, R.; Jata, K.V. Effects of pitting corrosion on the fatigue behavior of aluminum alloy 7075-T6: Modeling and experimental studies. *Mater. Sci. Eng. A* **2001**, *297*, 223–229. [CrossRef]

18. Shi, P.; Mahadevan, S. Damage tolerance approach for probabilistic pitting corrosion fatigue life prediction. *Eng. Fract. Mech.* **2001**, *68*, 1493–1507. [CrossRef]

19. Jakubowski, M. Influence of pitting corrosion on fatigue and corrosion fatigue of ship and offshore structures, Part II—Pit—Crack Interaction. *Pol. Marit. Res.* **2015**, *22*, 57–66. [CrossRef]

20. Genel, K. The effect of pitting on the bending fatigue performance of high-strength aluminum alloy. *Scr. Mater.* **2007**, *57*, 297–300. [CrossRef]

21. Azar, V.; Hashemi, B.; Rezaeeyazdi, M. The effect of shot peening on fatigue and corrosion behavior of 316L stainless steel in Ringer's solution. *Surf. Coat. Technol.* **2010**, *204*, 3546–3551. [CrossRef]

22. Ribeiro, J.; Monteiro, J.; Lopes, H.; Vaz, M. Moiré Interferometry assessement of residual stress variation in depth on a shot peened surface. *Strain* **2011**, *47*, e542–e550. [CrossRef]

23. Tič, V.; Lovrec, D. Development of linear servo hydraulic drive for material testing. In *International Conference "New Technologies, Development and Applications"*; Lecture Notes in Networks and Systems; Springer: Sarajevo, Bosnia and Herzegovina, 2020; Volume 128, pp. 104–113.

24. Muminovic, A.J.; Saric, I.; Repcic, N. Numerical analysis of stress concentration factors. *Proc. Eng.* **2015**, *100*, 707–713. [CrossRef]

25. Yan, W.; Guan, L.; Xu, Y.; Deng, J.-G. Numerical simulation of the double pits stress concentration in a curved casing inner surface. *Adv. Mech. Eng.* **2017**, *9*, 1–7. [CrossRef]

26. Pederson, N.L. Stress concentrations in keyways and optimization of keyway des. *J. Strain Anal. Eng. Des.* **2010**, *45*, 593–604. [CrossRef]

27. Sukumar, N.; Slorovitz, D.J. Finite element-based model for crack propagation in polycrystalline materials. *Comput. Appl. Math.* **2004**, *23*, 363–380.

28. Baydoun, M. Crack propagation criteria in three dimensions using the XFEM and an explicit–implicit crack description. *Int. J. Fract.* **2012**, *178*, 51–71. [CrossRef]

29. Fu, P.; Johnson, S.M.; Settgast, R.R.; Carrigan, C.R. Generalized displacement correlation method for estimating stress intensity factors. *Eng. Fract. Mech.* **2012**, *88*, 90–107. [CrossRef]

30. Ashoaibi, A.A.; Ariffin, A.K. Evaluation of stress intensity factor using displacement correlation techniques. *J. Kejuruter.* **2008**, *20*, 75–81. [CrossRef]

31. Newman, J.C., Jr.; Wu, X.R.; Venneri, S.; Li, C. *Small-Crack Effects in High-Strength Aluminum Alloys*; NASA RP-1309; Langley Research Center: Hampton, VA, USA, 1994.

32. Newman, J.C.; Phillips, E.P.; Everett, R.A. *Analysis of Fatigue and Fatigue Crack Growth under Constant-and Variable Amplitude Loading*; NASA/TM-1999–209329; Langley Research Center: Hampton, VA, USA, 1999.

33. Newman, J.C., Jr. Fatigue and crack-growth analyses under gigacycle loading on aluminium alloys. *Proc. Eng.* **2015**, *101*, 339–346. [CrossRef]

34. Huang, Z.; Wagner, D.; Bathias, C.; Paris, P.C. Subsurface crack initiation and propagation mechanisms in gigacycle fatigue. *Acta Mater.* **2010**, *58*, 6046–6054. [CrossRef]

35. Wang, Q.; Kashif-Khan, M.; Bathias, C. Current understanding of ultra-high cycle fatigue. *Theor. Appl. Mech. Lett.* **2012**, *2*, 031002. [CrossRef]

36. Wang, Q.T.; Li, T.; Zeng, X.G. Gigacycle fatigue behavior of high strength aluminum alloys. *Proc. Eng.* **2010**, *2*, 65–70. [CrossRef]

Article

Effect of Asymmetric Material Flow on the Microstructure and Mechanical Properties of 5A06 Al-Alloy Welded Joint by VPPA Welding

Zhaoyang Yan [1], Shujun Chen [1], Fan Jiang [1,*], Xing Zheng [2], Ooi Tian [3], Wei Cheng [4] and Xinqiang Ma [4]

[1] Engineering Research Centre of Advanced Manufacturing Technology for Automotive Components Ministry of Education, Faculty of Materials and Manufacturing, Beijing University of Technology, Beijing 100124, China; zhygyan@126.com (Z.Y.); sjche@bjut.edu.cn (S.C.)
[2] China Academy of Space Technology Department of Spacecraft Environmental Engineering, Beijing 100094, China; xingzheng_welder@163.com
[3] Department of Chemical and Materials Engineering, University of Alberta, Edmonton, AB T6G 1H9, Canada; Tooi@alberta.ca
[4] Laser Institute, Qilu University of Technology (Shandong Academy of Sciences), Jinan 250014, China; chengweijob@163.com (W.C.); maxin-qiang@163.com (X.M.)
* Correspondence: jiangfan@bjut.edu.cn; Tel.: +86-137-2008-7645

Abstract: The microstructure, texture, and mechanical properties of the asymmetric welded joint in variable polarity plasma arc (VPPA) welding were studied and discussed in this paper. The asymmetric welded joint was obtained through horizontal welding, where the effect of gravity caused asymmetric material flow. The results showed that the grain size and low angle grain boundary (LAGB) at both sides of the obtained welded joint were asymmetric; the grain size differed by a factor of 1.3. The average grain size of the Base Metal (BM), Lower-weld zone (WZ) and Upper-WZ were 25.73 ± 1.25, 37.87 ± 1.89 and 49.92 ± 2.49 μm, respectively. There is discrepancy between the main textures in both sides of the welded joint. However, the effect of asymmetric metal flow on the weld texture was not significant. The micro-hardness distribution was inhomogeneous, the lowest hardness was observed in regions with larger grain sizes and smaller low angle grain boundary. During tensile strength tests, the specimens fractured at the position with the lowest hardness although it has reached 89.2% of the strength of the BM. Furthermore, the effect of asymmetric metal flow and underlying causes of asymmetric weld properties in VPPA horizontal welding have been discussed and analyzed.

Keywords: microstructure; material flow; asymmetric weld; tensile strength; VPPA

Citation: Yan, Z.; Chen, S.; Jiang, F.; Zheng, X.; Tian, O.; Cheng, W.; Ma, X. Effect of Asymmetric Material Flow on the Microstructure and Mechanical Properties of 5A06 Al-Alloy Welded Joint by VPPA Welding. *Metals* **2021**, *11*, 120. https://doi.org/10.3390/met11010120

Received: 6 December 2020
Accepted: 6 January 2021
Published: 9 January 2021

Publisher's Note: MDPI stays neutral with regard to jurisdictional claims in published maps and institutional affiliations.

1. Introduction

Aluminum alloy as a high-performance metal for producing lightweight equipment is widely used in manufacturing industries to promote energy conservation, emission reduction and highly efficient production [1,2]. Variable Polarity Plasma Arc (VPPA) welding with advantages of high energy beam and variable polarity gas tungsten arc is known as a "zero-defect" process in the medium thickness welding of Al-alloys [3,4]. VPPA vertical-up welding is the most widely used method: gravity assists material flow into the keyhole pool, producing high-quality welded joints. However, in non-vertical-up welding gravity will cause asymmetric molten flow [5–7], resulting in an asymmetric weld which can weaken the weld quality. Variable position welding is the bottleneck of VPPA Al alloy welding; the flow of molten pool will be asymmetric because of the change of the angle between gravity direction and welding direction, resulting in the asymmetric thermal field distribution, thus affecting the mechanical properties of welding joints [2,6,7].

Recently, VPPA welding has caught wide attention in manufacturing industries [8–10]. In VPPA welding, an axisymmetric high-speed plasma flow can be created after being com-

111

pressed by circular water-cooled nozzle in which the shrinkage of the tungsten electrode in the circular nozzle is 3.5–4.5 mm [6–12]. A keyhole molten pool can be formed due to the high energy density of VPPA. The welded joint of VPPA welding with a low distortion rate is benefited by the concentrated heat input and great temperature gradient [13,14]. Keyhole evolution and the metal flow of the weld pool in plasma arc welding have been studied, indicating the high-speed plasma flow benefits weld formation at the rear side [15]. The published works prove the advantages of VPPA in aluminum alloy welding. However, non-vertical-up welding of VPPA should get more attention to widen the application of this technology.

An arc with reduced arc pressure while maintaining the ability to make a stable keyhole in VPPA horizontal welding has been proposed and verified in aluminum plates [6,7]. The optimal welding method and parameters of VPPA flat welding have been analyzed by observing the metal flow using X-ray system and measuring the arc pressure [16,17]. In previous researches, the weld formation and performance during VPPA horizontal welding has been studied by monitoring filler wire positions and torch angles [8,11]. However, the microstructure, texture distribution, and mechanical behavior of non-vertical-up welding have not been studied previously. As the weld quality and properties are the key to evaluating its application value, it is important to understand asymmetric metal flow under the scheme of non-vertical-up welding.

The welded joint quality produced by arc welding depends primarily on accurate positioning of the weld pool [3]. Exploring the microstructure and mechanical properties of welded joints in non-vertical-up welding of VPPA is essential for applying this advanced technology into the manufacture of modern components. Therefore, the texture distribution and grain behavior of the weld in horizontal welding are studied. This work aims to investigate the effect of asymmetric metal flow on microstructure, texture and mechanical properties of the weld of VPPA welding. By studying the difference of the microstructure and mechanical properties of asymmetric welded joints, the influence of asymmetric molten flow on the weld formation can be clarified in this paper, providing the theoretical basis for improving the weld quality of variable position welding.

2. Materials and Methods

2.1. Materials

5xxx Al-alloys with magnesium principal alloying element give substantial increases in good corrosion resistance, machinability and weldability [12–14]. It is one of the mostly widely used materials for aerospace, automotive and other industries. 5A06 Al plates with the dimension of 300 mm × 150 mm × 5 mm are selected as the base material in this paper, and the compositions were provided by Weihai Jindi Non-ferrous Metals Co., Ltd. (Weihai, China) [12]. To avoid evitable error during or after the welding process, the wire ER5183 with the diameter of 1.2 mm and the same chemical composition as the base materials is used [11], and the nominal chemical compositions are shown in Table 1.

Table 1. Chemical composition of 5A06 and ER5183 (wt.%) [11,12].

Materials	Mg%	Mn%	Cr%	Fe%	Si%	Zn%	Cu%	Al%
5A06 plate	5.80~6.80	0.50~0.80	-	<0.40	≤0.40	≤0.20	≤0.10	Bal.
ER5183	4.30~5.00	0.50~1.00	≤0.1	≤0.40	≤0.40	≤0.25	≤0.10	Bal.

2.2. Experimental Method and Procedure

VPPA welding in horizontal way was conducted to investigate the effect of asymmetric metal flow on the microstructure of Al welded joints. The welding process is shown in Figure 1, the Z axis is the direction of gravity. Generally, the welding direction adopted by VPPA is vertical-up. In vertical-up welding, gravity results in symmetric metal flow on both sides of the keyhole. There is no asymmetry in the formation and mechanical properties of welded joints [3]. In horizontal welding, gravity pushes most of the melted metals to the lower side, resulting in uneven flow of the molten pool. This results in an asymmetric weld and microstructures in the welded joint with poor mechanical properties. To study the effect of asymmetric metal flow on the microstructure and mechanical properties of the welded joint, several steps has been carried out such as acid pickling and oxide removal to eliminate the influence of external factors on the welded joint before VPPA welding [12]. A series of orthogonal tests were performed, a group of test parameters with beautiful appearance and good formation of weld were selected. According to the determined welding parameters, five tests were carried out, and a welding seam was selected to carry out mechanical properties tests after determining the stability of the welding. The main welding parameters including welding current (Direct Current Electrode Negative: DCEN; Direct Current Electrode Positive: DCEP), travel velocity and gas flow rate are shown in Table 2.

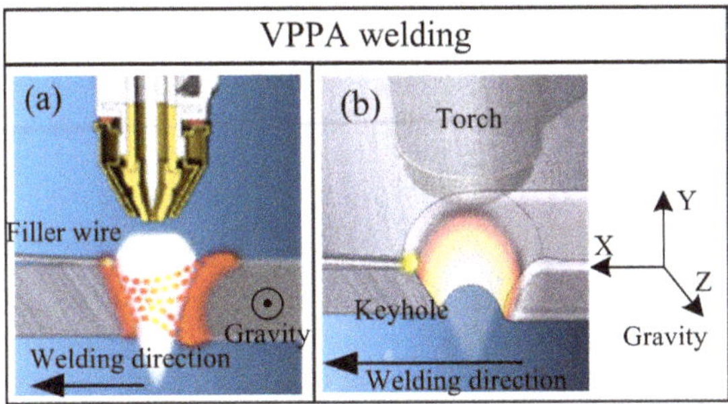

Figure 1. Schematic drawing of the VPPA welding process, (**a**) cross-section, (**b**) keyhole.

Table 2. Welding parameters.

Parameters		Value/Units
Welding currents	DCEN	135 A
	DCEP	165 A
Welding direction		Horizontal welding
Travel velocity		0.14 m/min
Welding power		≈4.5 KW
Wire feed speed		0.8 m/min
Wire diameter		1.2 mm
Shielding gas flow rate		Pure argon with 15 L/min
Plasma gas flow rate		Pure argon with 3.0 L/min

To study the effect of asymmetric metal flow on the microstructure and mechanical properties of welded joints, an optical microscope (OM, Olympus LEXT OLS4100, Toyko, Japan) is used to capture pictures of the cross-section of welded joint. Electron Back-Scattered Diffraction (EBSD) (SEM, JSM-7900F, JEOL, Toyko, Japan) is carried out to further study the microstructure, and the weld sample is electro-polished in a solvent consisting of 30% nitric acid (Jiang Yun, Kunming, China) and 70% methanol (Jiang Yun, Kunming, China) for 60 s at a temperature and voltage of −14 °C and 12 V [14,18]. Micro-hardness and tensile strength are used to evaluate the mechanical properties. The micro-hardness is measured with Vicker's method using a Vicker hardness tester (SCTMC DHV-1000 Hz, Hangzhou Quantum Testing Instrument Co., Ltd., Shanghai, China) with a 200 g (0.2 HV, 1.96 N) load for 15 s (the sampling points are shown in Figure 2a). There are four test lines, the distance between each test line is 1 mm, and the distance between each test point is 0.25 mm. Tensile tests were carried out at room temperature using a universal testing machine INSTRON-5569 (Norwood, MA, USA) under standard NoGBT228-2002 of P.R. China [14]. The size of the dog-bone specimen used to measure the tensile strength is shown in Figure 2b. The results are taken as the average value of three samples.

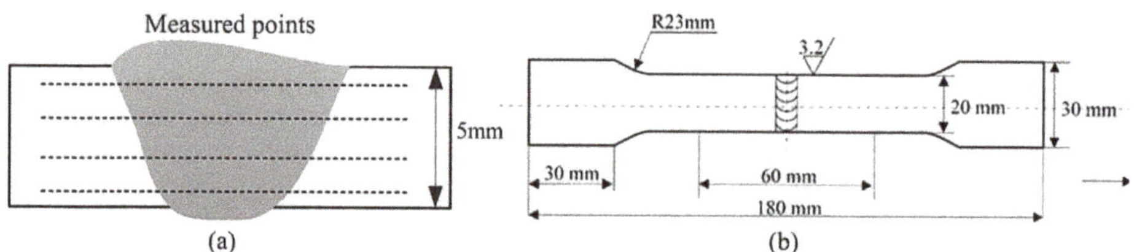

Figure 2. Schematic diagram for testing, (**a**) micro-hardness, (**b**) tensile strength.

3. Results

3.1. Weld Appearance

Figure 3 is the weld appearance and weld cross-section. The weld is smooth with high visual quality of the weld surface, as shown in Figure 3a. Generally, the "striae" after solidification during VPPA welding process is perpendicular to welding direction [8,11], however, in this case, the average angle between the "striae" and WD was about 45 degrees, as shown in Figure 3b. This phenomenon was caused by the asymmetric metal flow during horizontal welding. Gravity pushes most of the melted metals to the lower side (the side where the direction of gravity points to the side), resulting in uneven flow of the molten pool and asymmetric weld appearance. The obtained welded joint was asymmetric and the width of the weld surface and root surface were 11.5 and 7.9 mm, respectively. The deviation between the top and bottom center lines was 1.8 mm, as shown in Figure 3c. The deviation mainly happened at the weld surface, as shown in Figure 3c; it was indicated that gravity had little effect on the material flow on the exit side of the keyhole.

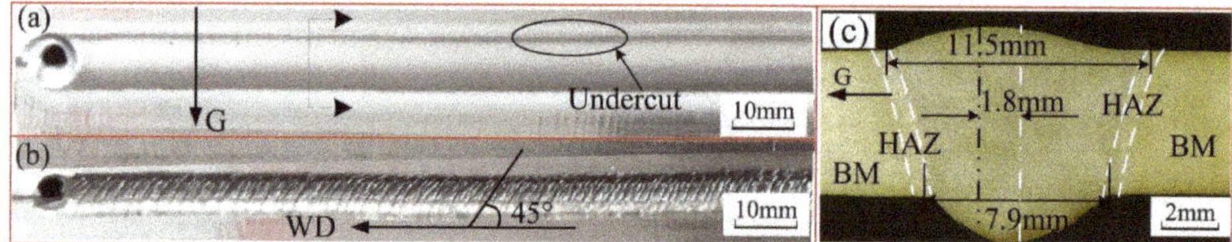

Figure 3. Weld appearance and macrostructure, (**a**) weld surface; (**b**) weld root surface; (**c**) weld cross-section.

3.2. Microstructure of the Welded Joint

The welded joint was characterized into base metal (BM), weld zone (WZ) and fusion line (FL). The fusion lines can be divided into Lower-FL (the side where the gravity points to the side, in which more melted metal gathers due to the effect of gravity) and Upper-FL (the opposite side). EBSD orientation mapping was carried out on those four regions, and the selected zones are shown in Figure 4a. The heat affected zones (HAZ) are included in the zones of c and e, as shown in Figure 4c,e. BM showed a typical rolling structure with elongated gain substructures, as shown in Figure 4b. Fine equiaxed grains appeared in the WZ, which is a typical microstructure of crystallization [19], as shown in Figure 4d. The grain structure from BM to WZ in Lower-FL showed rolling, columnar and equiaxed grain. The grain structure in Upper-FL is similar to that of Lower-FL, as shown in Figure 4c,e.

Figure 5 contains the corresponding inverse pole figure of the selected regions in Figure 4a and the corresponding color key in the contours on the right side. In this study, the transverse direction (TD) was parallel to the rolling direction of BM; normal direction (ND) represent the direction of the thickness of BM; and the welding direction (WD) was perpendicular to the rolling direction of BM. The density ranges (MUD) for BM, Lower-FL, WZ, and Upper-FL were 0.65 to 1.69, 0.60 to 1.50, 0.66 to 1.49, and 0.58 to 1.99, respectively. And the main textures were {101}, {111}, {102} and {001}, respectively. In general, the texture intensity can be decreased because of crystallization during melting and solidification, which is induced by shear deformation and high temperature. Compared to the maximum texture intensity in BM of 1.69, the maximum texture intensity of Lower-FL and WZ were 1.49 and 1.50, respectively. However, the maximum texture intensity in Upper-FL was 1.99, which was higher than that of the result in BM. The texture intensity difference between the tested regions was also affected by the materials. Therefore, there is no preferred distribution of grain orientation, which indicates that the effect of asymmetric metal flow on the texture distribution is not significant at the macro level.

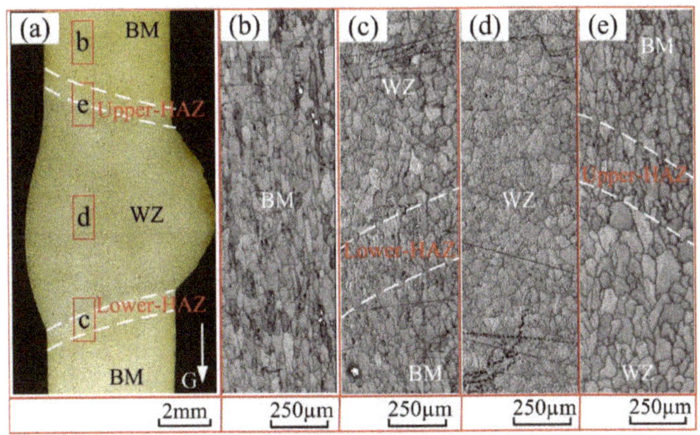

Figure 4. Macro- and Micro-structure of the welded joint, (**a**) welded joint; (**b**) BM; (**c**) Lower-FL; (**d**) WZ; (**e**) Upper-FL.

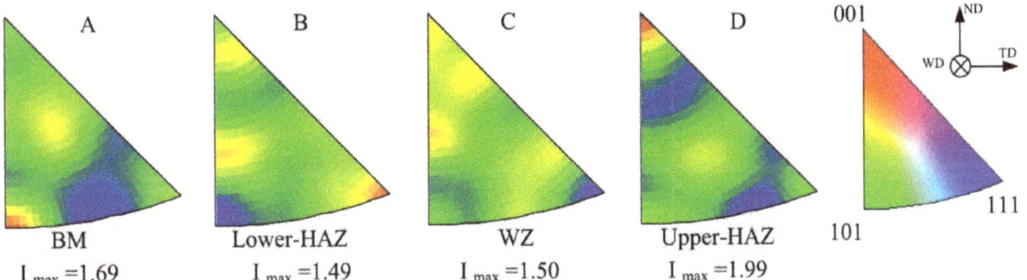

Figure 5. Inverse pole Figures, (**A**) BM; (**B**) Lower-HAZ; (**C**) WZ; (**D**) Upper-HAZ.

Figure 6 shows number fraction of the grain boundary angles of 2–15° and average grain size, the grain boundary angles between 2° and 15° were defined as low angle grain boundary (LAGB). The fraction of LAGBs decreased significantly from BM to WZ. LAGBs occupied 30.39%, 14.34%, 10.21% and 12.95% from zone b to zone e. LAGBs not only affect the plasticity, toughness, fracture and corrosion but also affect the brittleness and fatigue property of the material. In this case, the number fraction of LAGBs of the WZ reduced in comparison with the BM, indicating that significant recovery or recrystallization had occurred during welding process [20,21]. The material fluidity can be improved by heat input; it also can improve complete dynamic recrystallization [22]. The average grain sizes, which were calculated by the linear intercept method, of the selected regions were also shown in Figure 6. The average grain size in BM was 25.73 ± 1.25 μm, showing a typical rolling structure with elongated gain substructures, as shown in zone b. Fine equiaxed grains (a typical feature of dynamic recrystallization) appeared in WZ [8] with a grain size of 34.55 ± 1.72 μm, as shown in zone d. It was observed that the Upper-FL (zone d) has larger grain size than the Lower-FL (zone c) in the asymmetric welded joint. The average grain size in Lower-FL and Upper-FL were 34.33 ± 1.71 μm and 43.15 ± 2.16 μm. The average grain size, in decreasing order, is zone e (Upper-FL) > zone d (WZ) > zone c (Lower-FL) > zone b (BM). It can be concluded from this section that the average grain sizes at both side of the asymmetric welded joint were asymmetric. An asymmetric temperature distribution of the weld pool was produced by the asymmetric metal flow during the welding process, leading to the non-equilibrium nucleation of grains in the welded joint. That is, the asymmetric metal flow during VPPA horizontal welding has an influence on grain grown during the crystallization.

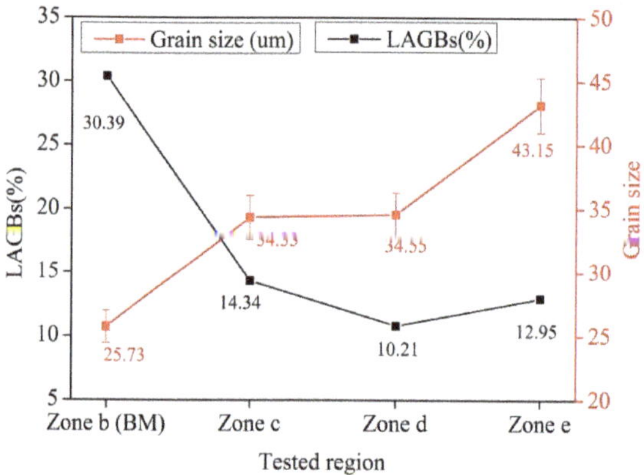

Figure 6. Number fraction of the LAGBs and grain size.

3.3. Microstructure and Texture Evolution at Both Side of the Weld

This section explores the effect of asymmetric metal flow on the microstructure, texture and mechanical properties of Al welded joint by VPPA welding. The microstructure and texture evolution on both sides of the welded joint were studied in this section, including four tested regions Lower-HAZ (heated affected zone where gravity points to the side), Lower-WZ, Upper-WZ and Upper-HAZ, as shown in Figure 7a. The corresponding band contrast results are shown in Figure 7b–e, and the corresponding grain boundaries are shown in Figure 7(b1–e1). The grain in Lower-HAZ was columnar crystal, while the grain size in Upper-HAZ was inhomogeneous. Both the Lower-WZ and Upper-WZ belong to the perfect recrystallization zone, shown in the structure of equiaxed grains, whereas, the grain size in those zones are different; the average grain size in Upper-WZ is larger than that of the size in Lower-WZ. The results were caused by the inhomogeneous crystallization temperature influenced by the asymmetric metal flow during welding.

Figure 8 shows the inverse pole Figure (IPF) color maps and IPFs of the selected zone in Figure 7a. The texture of the Upper-HAZ and Upper-WZ was {001} while that of the Lower-HAZ was {111}. The highest texture of those zones in Lower-HAZ, Lower-WZ, Upper-WZ and Upper-HAZ were 4.40, 2.32, 3.32 and 2.92, respectively. It can be obtained that the texture intensity of the HAZ and WZ was higher than that of BM. That is, the heat treatment during the welding process increased the concentration of the grain orientation distribution. In horizontal welding, melted metal gathers at the side where gravity points to the side; the dynamic recrystallization at this region was more sufficient. The texture intensity of this region was higher than that of the opposite side. Whereas, the texture intensity of those regions were not significantly different, as the range was between 2.3 and 4.4. It can be obtained that asymmetric metal flow has little effect on the texture distribution.

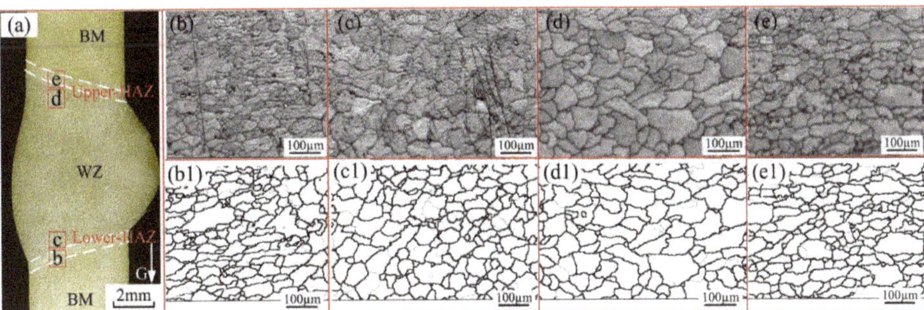

Figure 7. Microstructures at both side of the weld, (**a**) welded joint; (**b**–**e**) microstructure of the selected regions in (**a**); (**b1**–**e1**) grain boundary of (**b**–**e**).

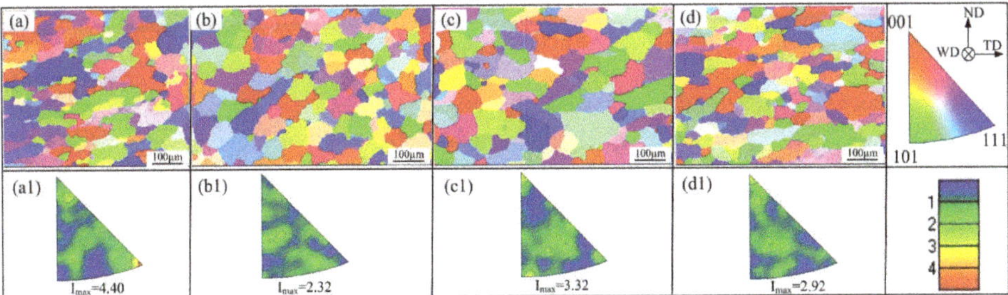

Figure 8. Inverse pole Figure color maps, (**a**–**d**) the selected zone shown in Figure 7a, (**a**) zone b; (**b**) zone c; (**c**) zone d; (**d**) zone e; (**a1**–**d1**) the corresponding IPF of (**a**–**d**).

The average grain size and LAGB distribution at the selected zones are shown in Figure 9. The average grain size in BM was 25.73 ± 1.25 µm, which is the smallest grain in the whole welded joint. And the average grain size in the Upper-WZ was about 49.92 ± 2.49 µm, which is the biggest grain in the welded joint. The average grain sizes in Lower-HAZ and Upper-HAZ were 30.79 ± 1.54 and 39.11 ± 1.96 µm, respectively. The average grain size at the side where gravity points to the side was larger than the opposite side because the heat accumulation at the opposite reduced during welding process. Lots of heat was taken to the lower side with the melted metal flow [2]. A high heating rate is beneficial for accelerating the crystallization [14]. Meanwhile, the crystallization in the lower side was more complete than that of the opposite side. The misorientation angle distribution was shown in the black line on Figure 9. The fraction of LAGBs decreased significantly from BM to Upper-HAZ. LAGBs occupied 30.39%, 16.53%, 10.08%, 10.21%, 13.01% and 12.56%, respectively. LAGBs are primarily caused by dynamic recovery during deformation, and are mainly within large grains or along grain boundaries [14]. In this case, the range of LAGBs in WZ was between 10.08 and 13.01%, indicating that the difference of the crystallization in both sides of the welded joint was not noticeable. Whereas, a significant difference was obtained between Lower-HAZ and Upper-HAZ, indicating that material flow and thermal process were different on both sides of the HAZ. The LAGBs distribution in the whole welded joint was within 10–13%, which is acceptable within the margin of error.

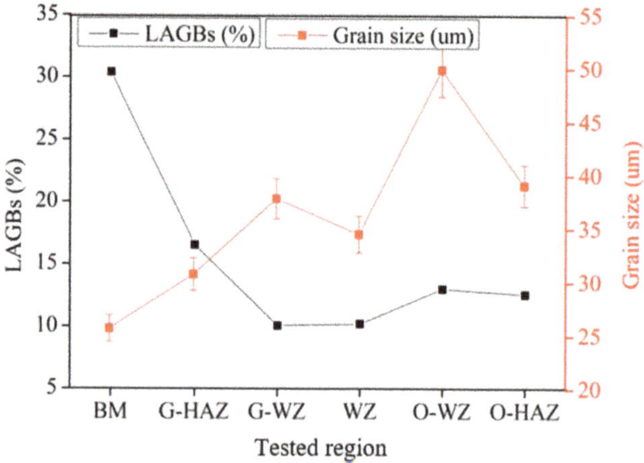

Figure 9. Number fraction of the LAGBs and grain size.

3.4. Mechanical Properties

The micro-hardness distribution of the welded joint is shown in Figure 10. Generally, the micro-hardness in the WZ, FL and HAZ are softer than that of BM in VPPA weld, and the micro-hardness nearby the FL is softer than that of the WZ and HAZ, similar to several researches [23,24]. For an overall view of the welded joint in this paper, the highest micro-hardness appeared in BM, the value was among 85 HV and 90 HV. The micro-hardness of WZ was lower than that of BM, the value was among 75 HV and 85 HV. The lowest micro-hardness appeared nearby the FL, but the value of the Upper-FL was lower than that of the Lower-FL; the micro-hardness of Lower-FL was about 70–75 HV, and that of Upper-FL was about 65–70 HV. For an analysis of the hardness, the lowest micro-hardness appeared near the Upper-FL, as shown in Figure 4. The asymmetric metal flow contributed to the inhomogeneous hardness distribution.

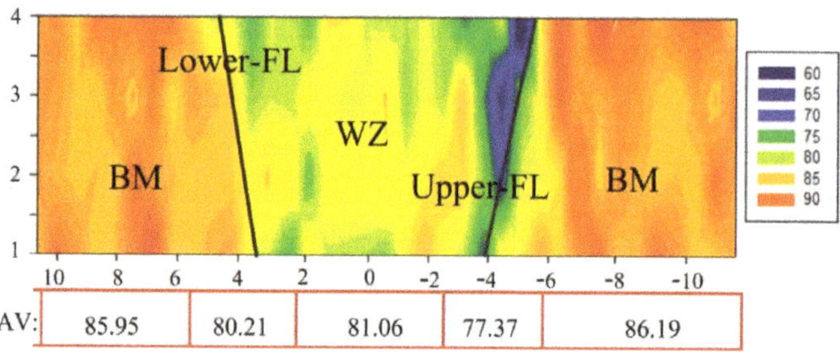

Figure 10. Microhardness distribution of the welded joint.

According to the relation between the tensile strength and micro-hardness, the tensile strength is proportional to the micro-hardness at the same welds [23]. The tensile strength and elongation of the asymmetric welded joint and BM are shown in Table 3. The results are the average value of two different tests of the same materials. The weld was tested to 319.8 ± 11.2 MPa, reaching about 89.2% of that of the BM. The value was the average value of three tests, and the specimens were all fractured at the region nearby the Lower-FL. For elongation, one of the important indexes to judge material properties, the result of the weld was about $11.4 \pm 0.8\%$, reaching about 45.2% of the BM. Figure 11 is the SEM fractography of the fractured surfaces. Many tears and dimples appeared in the fracture surface of the BM, showing a quasi-cleavage fracture, which was one of the ductile fractures, as shown in Figure 11a,b. As for the weld, the fractured surface has the characteristics of ductile fracture-equiaxial dimples and elongation dimples, as shown in Figure 11c,d. It is worth noting that the welds fractured at the Upper-FL, indicating that the Upper-FL is the weakest link in the whole welded joint. The mechanical properties of the welded joint were significantly affected by the asymmetric metal flow during VPPA horizontal welding.

Table 3. Tensile strength of the welded joint and BM.

Tested Specimen	Fractured Position	Tensile Strength/MPa	Elongation/%
Base metal	-	358.7 ± 9.3	25.2 ± 1.4
Welds	Upper-FL	319.8 ± 11.2	11.4 ± 0.8

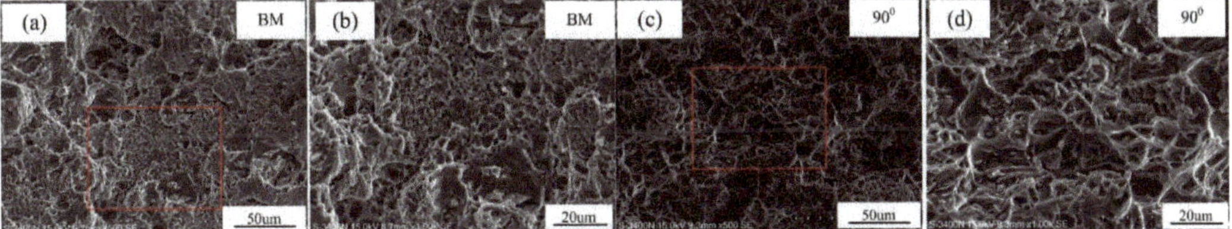

Figure 11. SEM of the fractured surfaces, (**a,b**) BM; (**c,d**) weld.

4. Discussion

From the above results, the microstructure, texture evolution and mechanical properties of the asymmetric welded joint have been obtained. The weakest zone of the welded joint appeared nearby the Upper-FL. The effect of asymmetric metal flow and underlying causes of asymmetric weld properties in VPPA horizontal welding have been discussed in this section. Figure 12 shows the keyhole molten pool of VPPA horizontal welding. There were few noticeable asymmetric phenomena in the keyhole entrance, as shown in Figure 12a. The metal was melted at the leading side and solidified at the rear side after flow around the keyhole, forming a surface weld. Figure 12b is the keyhole exit side, which is an irregular ellipse. In VPPA horizontal welding, lots of the melted metal flows to the side where gravity points, while a little stays at the opposite side [2,8]; the melting layer at the lower side was 1.9 mm, whereas, at the opposite side it was only 1.1 mm. The keyhole shapes and weld profiles are shown in Figure 13. With the accumulation of the welding process, a portion of the melted metal at the opposite side flowed to the lower side, forming an asymmetric keyhole, as shown in Figure 13a, resulting in an inhomogeneous keyhole molten pool, as shown in Figure 13c. It's worth noting that the melted metal has a high temperature, that is, with the accumulation of the melted metal, the temperature distribution around the keyhole is also inhomogeneous. The "striae" can express the situation of solidification during the welding process, as shown in Figure 13b. The heat emission condition of the keyhole is consistent because of the same material, whereas, the initiation temperature at both sides of the pool is uneven. The temperature of the side where gravity points to the side is higher than that of the opposite side [2].

For the free-growth model, the grain size is inversely proportional to the undercooling ΔT; that is, grain grow depends only on undercooling (initial temperature) in this case. The side with high temperature obtains small grain size, which agrees with the results, as shown in Figure 9. According to the Hall-Petch relationship, the hardness increases as the grain size decreases [25], which also agrees with this work, as shown in Figure 10. The region nearby the Upper-FL has the lowest hardness distribution, and the grain size here was about 49.92 μm, which is the highest value throughout the whole welded joint. The lower hardness distribution at the Upper-FL may weaken the tensile strength. For the mechanical properties of a material, the smaller the grains, the more uniform the plastic deformation of the material is, because the plastic deformation can be dispersed among more grains. The smaller grains can enhance the dislocation among the grains, resulting in better strength and toughness. That is, materials with larger grain size have poor mechanical properties [22,26]. All this agrees with the results in this paper.

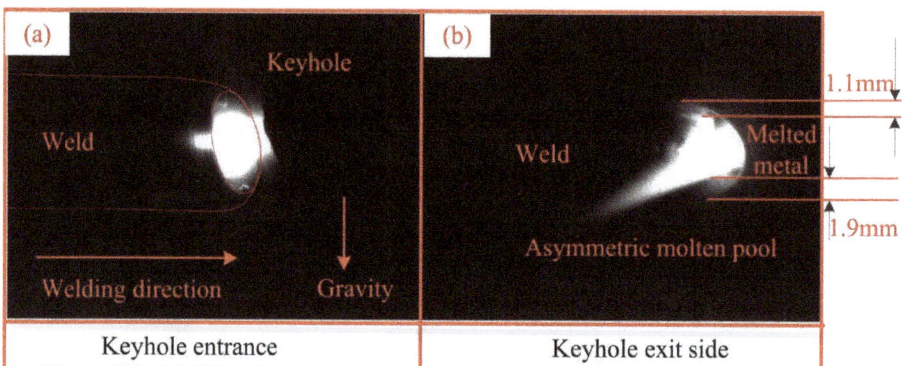

Figure 12. Keyhole molten pool of VPPA horizontal welding, (**a**) keyhole entrance, (**b**) keyhole exit.

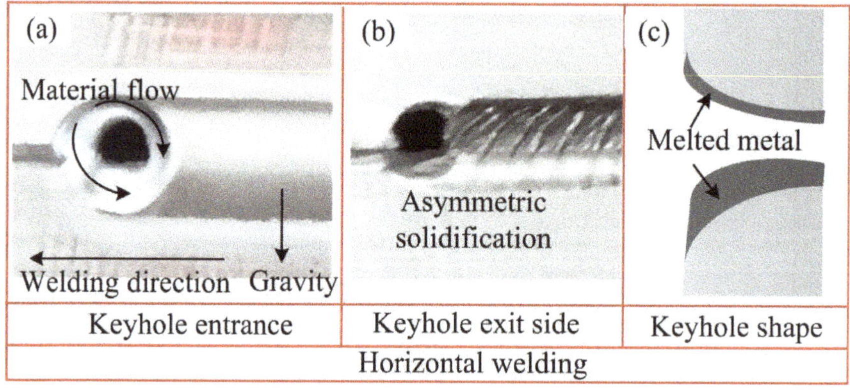

Figure 13. Material flow and weld formation of VPPA horizontal welding, (**a**) keyhole entrance, (**b**) keyhole exit, (**c**) welded joint.

Generally, the energy of LAGBs is mainly derived from dislocation, with the density of dislocation depending on the dislocation difference between grains. The interface energy of HAGBs is higher than that of LAGBs [11]. LAGBs can prevent the crack from expanding more. The blocking effect in the grain boundary is affected by the interface energy: the smaller the energy of the grain boundary, the better its stability, and the clearer the blocking effect [12]. Thus, materials tend to fracture with high probability at the region with higher HAGBs, which is in accordance with our tensile test results. The texture component distribution in the welded joint also has an influence on properties. However, the results obtained in this paper show that the effect of the asymmetric metal flow of horizontal welding on the texture distribution is not noticeable, as stated above.

In VPPA horizontal welding, the melted metal around the keyhole flowed to the side where gravity points to the side along the keyhole boundary, causing an asymmetric metal distribution on both sides of the molten pool [2,11]. The temperature distribution around the pool was also affected by the asymmetric metal flow, which may directly affect the weld solidification process [14]. The crystallization process at both sides of the weld can be affected by the asymmetric metal flow during the welding process, resulting in an asymmetric grain size and grain-boundary angle distribution and poor mechanical properties. It is worth noting that all the mechanical properties lost from the weld of VPPA horizontal welding were lost directly due to gravity. The gravity effects on the asymmetric metal flow should be weakened to improve the mechanical properties of the asymmetric welded joint.

5. Conclusions

The effect of asymmetric material flow on the microstructure, texture and mechanical properties of Al welded joint by VPPA welding has been studied in this paper. The results obtained are as follows:

(1) In VPPA horizontal welding, asymmetric metal flow created an asymmetric weld, resulting in asymmetric grain size, texture and GBA distribution on both sides of the welded joint.
(2) The density ranges of texture in WZ, HAZ and BM were within 1.99, indicating the effect of asymmetric metal flow on the texture distribution was not significant.
(3) The largest grain size appeared in the Upper-WZ, the average grain size was about 49.92 ± 2.49 μm, which was about 1.3 times than the opposite side.
(4) The hardness near the Upper-FL was lower than that of the Lower-FL. A higher density of HAGB and lower hardness at the Upper-FL weakened the mechanical properties of the welded joint, causing the tensile specimens to break at this region.

Author Contributions: Conceptualization, Z.Y. and F.J.; methodology, S.C.; validation, Z.Y., O.T. and X.Z.; investigation, Z.Y.; resources, X.Z. and X.M.; data curation, Z.Y. and W.C.; writing—original draft preparation, Z.Y.; writing—review and editing, F.J.; visualization, F.J. and S.C.; supervision, S.C.; funding acquisition, F.J. All authors have read and agreed to the published version of the manuscript.

Funding: This work is supported by National Natural Science Foundation of China (U1937207), R&D projects in key areas of Guangdong province (2018B090906004) and Major Science and Technology Innovation Project of Shandong Province (2019JZZY010452).

Institutional Review Board Statement: Not applicable.

Informed Consent Statement: Not applicable.

Data Availability Statement: Not applicable.

Conflicts of Interest: The authors declare no conflict of interest.

References

1. Wang, H.; Liu, X.; Liu, L. Research on Laser-TIG Hybrid Welding of 6061-T6 Aluminum Alloys Joint and Post Heat Treatment. *Metals* **2020**, *10*, 130. [CrossRef]
2. Yan, Z.Y.; Chen, S.J.; Jiang, F.; Zhang, W.; Tian, O.; Huang, N.; Zhang, S.L. Weld properties and residual stresses of VPPA Al welds at varying welding positions. *J. Mater. Res. Technol.* **2020**, *9*, 2892–2902. [CrossRef]
3. Kartal, M.E.; Liljedahl, C.D.M.; Gungor, S.; Edwards, L.; Fitzpatrick, M.E. Determination of the profile of the complete residual stress tensor in a VPPA weld using the multi-axial contour method. *Acta Mater.* **2008**, *56*, 4417–4428. [CrossRef]
4. Liu, Z.M.; Fang, Y.X.; Cui, S.L.; Luo, Z.; Liu, W.D.; Liu, Z.Y.; Jiang, Q.; Yi, S. Stable keyhole welding process with K-TIG. *J. Mater. Process. Technol.* **2016**, *238*, 65–72. [CrossRef]
5. Liu, Z.M.; Cui, S.L.; Luo, Z.; Zhang, C.Z.; Wang, Z.M.; Zhang, Y.C. Plasma arc welding: Process variants and its recent developments of sensing, controlling and modelling. *J. Manuf. Process.* **2016**, *23*, 315–327. [CrossRef]
6. Zhang, Q.L.; Yang, C.L.; Lin, S.B.; Fan, C.L. Novel soft variable polarity plasma arc and its influence on keyhole in horizontal welding of aluminium alloys. *Sci. Technol. Weld. Join.* **2014**, *6*, 493–499. [CrossRef]
7. Zhang, Q.L.; Yang, C.L.; Lin, S.B.; Fan, C.L. Soft variable polarity plasma arc horizontal welding technology and weld asymmetry. *Sci. Technol. Weld. Join.* **2015**, *4*, 296–306. [CrossRef]
8. Yan, Z.Y.; Chen, S.J.; Jiang, F.; Zhang, W.; Huang, N.; Chen, R. Control of gravity effects on weld porosity distribution during variable polarity plasma arc welding of aluminum alloys. *J. Mater. Process. Technol.* **2020**, *282*, 116693. [CrossRef]
9. Vyskoč, M.; Sahul, M.; Dománková, M.; Jurči, P.; Sahul, M.; Vyskočová, M.; Martinkovič, M. The Effect of Process Parameters on the Microstructure and Mechanical Properties of AW5083 Aluminum Laser Weld Joints. *Metals* **2020**, *10*, 1443. [CrossRef]
10. Yang, B.; Tan, C.W.; Zhao, Y.B.; Wu, L.J.; Chen, B.; Song, X.G.; Zhang, H.; Feng, J. Influence of ultrasonic peening on microstructure and surface performance of laser-arc hybrid welded 5A06 aluminum alloy joint. *J. Mater. Res. Technol.* **2020**, *9*, 9576–9587. [CrossRef]

11. Kalenda, M.; Madeleine, D.T. Corrosion fatigue behaviour of aluminium alloy 6061-T651 welded using fully automatic gas metal arc welding and ER5183 filler alloy. *Int. J. Fatigue* **2011**, *33*, 1539–1547.
12. Yan, Z.Y.; Chen, S.J.; Jiang, F.; Huang, N.; Zhang, S.L. Material flow in variable polarity plasma arc keyhole welding of aluminum alloy. *J. Manuf. Process.* **2018**, *36*, 480–486. [CrossRef]
13. Tian, Z.; Zhou, Q.; Du, Y.; Ma, Y.; Wang, W. Study of weld process and distortion control on VPPA welded laminose aluminium alloy. *New Technol. New Process.* **2015**, *11*, 104–106.
14. Jiang, X.Q.; Chen, S.J.; Gong, J.L. Effect of non-axisymmetric arc on microstructure, texture and properties of variable polarity plasma arc welded 5A06 Al alloy. *Mater. Charact.* **2018**, *139*, 70–80. [CrossRef]
15. Li, T.Q.; Chen, L.; Zhang, Y.; Yang, X.M.; Lei, Y.C. Metal flow of weld pool and keyhole evolution in gas focusing plasma arc welding. *Int. J. Heat Mass Transf.* **2020**, *150*, 119296. [CrossRef]
16. Xu, B.; Chen, S.J.; Tashiro, S.; Fan, J.; Văn, A.N.; Tanaka, M. Material flow analyses of high-efficiency joint process in VPPA keyhole flat welding by X-ray transmission system. *J. Clean. Prod.* **2020**, *250*, 119450. [CrossRef]
17. Xu, B.; Chen, S.J.; Jiang, F.; Phan, H.L.; Tashiro, S.; Tanaka, M. The influence mechanism of variable polarity plasma arc pressure on flat keyhole welding stability. *J. Manuf. Process.* **2019**, *37*, 519–528. [CrossRef]
18. Ahmed, M.M.Z. The Development of Thick Section Welds and Ultra-Fine Grain Aluminum Using Friction Stir Welding and Processing. Ph.D. Thesis, The University of Sheffield, Shaffield, UK, 2009.
19. Monajati, H.; Zoghlami, M.; Tongne, A.; Jahazi, M. Assessing Microstructure-Local Mechanical Properties in Friction Stir Welded 6082-T6 Aluminium Alloy. *Metals* **2020**, *10*, 1244. [CrossRef]
20. Jazaeri, H.; Humphreys, F.J. The transition from discontinuous to continuous recrystallization in some aluminium alloys. *Acta Mater.* **2004**, *52*, 3239–3250. [CrossRef]
21. Chen, H.Y.; Fu, L.; Liang, P. Microstructure, texture and mechanical properties of friction stir welded butt joints of 2A97 Al Li alloy ultra-thin sheets. *J. Alloys Compd.* **2017**, *692*, 155–169. [CrossRef]
22. Song, Y.B.; Yang, X.Q.; Cui, L.; Hou, X.P.; Shen, Z.K.; Xu, Y. Defect features and mechanical properties of friction stir lap welded dissimilar AA2024–AA7075 aluminum alloy sheets. *Mater. Des.* **2014**, *55*, 9–18. [CrossRef]
23. Cahoon, J.R.; Broughton, W.H.; Kutzak, A.R. The determination of yield strength from hardness measurements. *Metall. Trans.* **1971**, *2*, 1979–1983.
24. Han, J.H.; Kim, D.Y. Determination of three-dimensional grain size distribution by linear intercept measurement. *Acta Mater.* **1998**, *46*, 2021–2028. [CrossRef]
25. Hemmesi, K.; Mallet, P.; Farajian, M. Numerical evaluation of surface welding residual stress behavior under multi axial mechanical loading and experimental validations. *Int. J. Mech. Sci.* **2020**, *168*, 105127. [CrossRef]
26. Sato, Y.S.; Urata, M.; Kokawa, H.; Ikeda, K. Hall-Petch relationship in friction stir welds of equal channel angular-pressed aluminium alloys. *Mater. Sci. Eng. A* **2003**, *354*, 298–305. [CrossRef]

Article

Research on Laser-TIG Hybrid Welding of 6061-T6 Aluminum Alloys Joint and Post Heat Treatment

Hongyang Wang *, Xiaohong Liu and Liming Liu

Key Laboratory of Liaoning Advanced Welding and Joining Technology & School of Materials Science and Engineering, Dalian University of Technology, Dalian 116024, China; Liuxh_72@163.com (X.L.); liulm@dlut.edu.cn (L.L.)
* Correspondence: wang-hy@dlut.edu.cn

Received: 22 November 2019; Accepted: 7 January 2020; Published: 15 January 2020

Abstract: The 6061-T6 aluminum (Al) alloys was joined by the laser induced tungsten inert gas (TIG) hybrid welding technique. It mainly studied the influences of welding parameters, solution, and aging (STA) treatment on the microstructure and tensile properties of Al alloy hybrid welding joints. Microstructures of the joints were also analyzed by optical microscopy and transmission electron microscopy. Results showed that the laser induced arc hybrid welding source changed the microstructure of the fusion zone and heat effect zone. STA treatment effectively improved the mechanical properties of the softening area in the hybrid welding joint, whose values of the tensile strength and elongation were on average 286 MPa and 20.5%. The distribution of the reinforcement phases and dislocations distributed were more uniform, which improved the property of STA treated joint.

Keywords: Al alloy; laser induced arc hybrid welding; heat treatment; microstructure

1. Introduction

Aluminum alloys are used as structural materials, due to the high strength, excellent corrosion resistance, and low density, which makes them promising alternatives to steels for lightweight structures such as vehicles and airplanes [1–7]. Aluminum alloys are also extensively used in aviation, vessel, and electronics industries due to their light weight, high specific strength, good corrosion resistance, and excellent mechanical properties [8–16]. Aluminum profile heat treatment is one of the most important processes in the production of materials. Compared with other processes, heat treatment generally does not change the shape of the specimen and the overall chemical composition, but rather provides or improves the performance of the specimen by changing the internal microstructure of the specimen, or the chemical composition of the surface of specimen. As a heat treatable alloy, 6061-T6 had found a wide application in the fabrication of lightweight structures. It was excellent weldability over other high strength aluminum alloys [17].

Lots of welding methods were used in the joining of aluminum alloys such as tungsten inert gas (TIG), metal inert gas (MIG), laser beam welding (LBM), friction stir welding (FSW), hybrid welding process, etc. Ahmad R et al. [18] investigated the effects of three filler wires, namely, ER5356, ER4043, ER4047, on the microstructures and properties of a butt welded joint of AA6061 aluminum alloy by TIG welding. Kulekci et al. [19] compared the mechanical properties of welded joints of AW-6061-T6 aluminum alloy gained by FSW and MIG, respectively. They pointed out that the joint obtained with the FSW welding process had better mechanical properties and a narrower heat affected zone than those obtained by the MIG process. Narsimgachary et al. [20] welded a 6061-T6 aluminum alloy by the laser welding process and studied the temperature profile on the microstructure and mechanical properties of the joint. Chen Zhang et al. [21] used 5–6 kW laser-arc hybrid welding source to weld an

AA6082 aluminum alloy, and pointed out that the hybrid welding had some advantages, compared with the joints welded by simple laser or pure arc, such as the higher fatigue limit, lower percent porosity, grains refinement, etc.

Since the 6061-T6 aluminum alloy can be strengthened by the heat treatment process, several researchers applied post-weld heat treatment processes to improve the properties of the welded joints. Mohammad et al. [22] studied the influence of solution treatment and subsequent artificial aging treatment on the joint of AA6061-T651 aluminum alloy by the TIG welding process. The results showed that post-weld heat treatment significantly improved the mechanical properties of the joint, due to grain refinement and precipitation hardening. Dong peng et al. [23] analyzed the effects of aging treatment and heat treatment on the microstructure and mechanical properties of TIG-welded 6061-T6 aluminum alloy joints, and they pointed out that the mechanical properties of the joints were improved significantly because of distributing of few precipitates in the welded seam. Elangovan and Balasubramanian [24] investigated the effect of various post-heat treatments, namely, solution treatment, aging treatment, and the combination of them on tensile properties of FSW-ed AA6061 aluminum alloy.

With the development of the Al alloy manufacture, the requirement of the Al welding joint was not only the tensile strength, but also the plastic deformation capacity. It was very important for the multilink manufacture process, when the welding process was in the midst link but not the final process. The tensile strength of the above welding joints was high enough, but the ductility of the joints still needs to be improved. The ductility of the joints was mainly influenced by the microstructure, precipitate, and the dislocation of the joint. The low power laser induced arc hybrid welding source changed the heat distribution of the fusion zone, which made an obvious effect on the microstructure of the joint. Therefore, the ductility of the Al joints could be improved in laser induced arc hybrid welding process. Therefore, the low power pulsed laser induced TIG hybrid welding source was used to weld a 6061-T6 aluminum alloy in this essay. It was made to explore the influence of hybrid welding parameters and PWHT treatment on the tensile strength and ductility of the hybrid welding Al alloy joint.

2. Experiments

Commercial rolled 3 mm thick 6061-T6 aluminum alloy plates were used as base metal (BM), and machined into the size of 100×50 mm. The ER5365 filler wire with diameter of 1.2 mm was used in the experiments. Their chemical compositions are shown in Table 1. Figure 1 is the optical microstructure of BM. It was found that the grains in BM were elongated along the rolling direction, the grey phase was the matrix of α (Al), and black and white precipitate phase were β-Mg_2Si and Al-Fe-Si, respectively. Table 2 presents the main mechanical properties of BM [2].

Table 1. Chemical compositions (wt%) of base metal and filler wire.

Elements	Mg	Si	Fe	Zn	Cu	Al
6061-T6 (BM)	1.00	0.55	0.36	0.01	0.26	Bal
ER5356 (filler wire)	5.10	0.20	0.20	0.10	0.01	Bal

Table 2. Mechanical properties of BM.

Material	Ultimate Tensile Strength (MPa)	Elongation (%)	Vickers Hardness (Hv)
6061-T6 (BM)	340	19	115

Figure 1. Microstructure of aluminum (Al) alloy BM.

The Al alloy butt joints are welded by a laser induced arc hybrid welding source and the schematic diagram of the welding process is shown in Figure 2. In the experiment, α is the angle between TIG touch and horizontal, β is the angle between filler wire and horizontal, d is the distance between the laser incident point and filler wire on the plate, and D_{la} is the distance between the laser incident point and tungsten electrode on the plate. The main welding parameters are shown in the Table 3. The pulse laser is used in the experiments. The pulse duration is 2 ms.

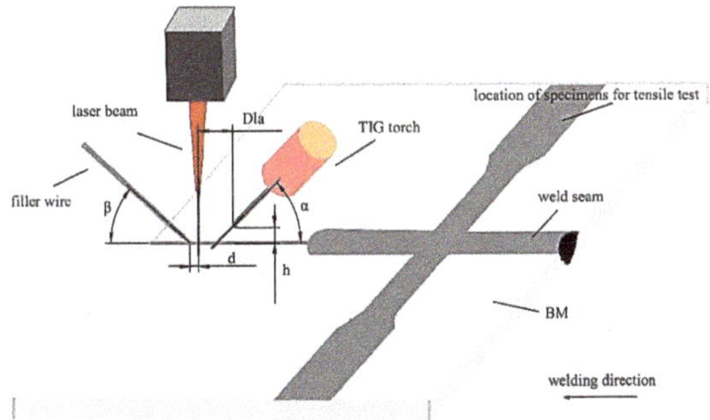

Figure 2. Laser induced arc hybrid welding process and the schematic diagram.

Table 3. Main parameters of laser induced arc welding Al alloy processes.

Arc Current (A)	Average Laser Power (W)	Welding Speed (mm·min⁻¹)	Pulse Repetition Rate (Hz)	Peak Power (W)
100–150	800–1000	500–700	40	6000~8500
β (°)	h (mm)	d (mm)	α (°)	D_{la} (mm)
50	1.5	1.5	45	2

To understand the influence of post-weld heat treatment on the mechanical properties of the joints, the specimens were divided into two categories, namely, hybrid welded (HW) joints and solution treated and aged (STA) joints. In the process of STA treatment, the joints were treated in the furnace at 475, 520, and 565 °C for 1 h, quenched in a cold water bath and then were kept in the induction furnace at 155 °C for 8 h. The property of the STA joint with 520 °C treatment showed the best results, thus only the STA joint with 520 °C treatment was discussed in the article.

To evaluate the mechanical properties of specimens, BM, HW, AG, and STA joints were machined into the required dimensions according to the ASTM E8 E8M-11 standard. The tensile test was carried out at room temperature and a speed of 2 mm/min by using an electro-mechanical controlled universal testing machine. Three samples were tested for each treated joints and BM, and the average values were calculated to evaluate the tensile properties of the specimens.

The joints were sliced into a small suitable size, grinded, and polished according to the metallographic specimen preparation standard, etched by Keller reagent (1.5 vol% HF + 1.5 vol% HCl + 2.5 vol% HNO_3 + 95 vol% H_2O) for 1 min, and then observed by an optical microscope (OM). Hardness test was performed on the cross section of the joint. The distance between each two test points is 0.25 mm. Scanning electron microscopy (SEM) was used to character the fracture surfaces of the tensile specimens and understand the failure patterns of the specimens. The phase compositions of welding joints fractures were identified by a PANayltical (Almelo, Netherlands) Empyrean powder X-ray diffractometer. The strengthening phases and dislocations were observed by transmission electron microscopy (TEM), and the TEM samples were prepared by double-jet electrolytic polishing in an electrolyte (30 vol% nitric acid and 70 vol% methanol) at a temperature of −35 °C with a voltage of 20 V.

3. Results

3.1. Welding Process

In order to investigate the influence of the main welding process parameters on the geometrical dimensions of the weld seam, the influence of the laser beam power, arc current, and welding speed on the welding joint morphology are shown in Figure 3.

Number	P I V (W) (A) (mm·min⁻¹)	Welding Joint Morphology	Cross Section of Joints
1	750 130 500		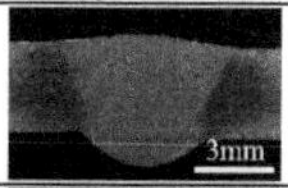
2	750 150 550		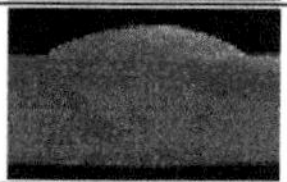
3	750 160 600		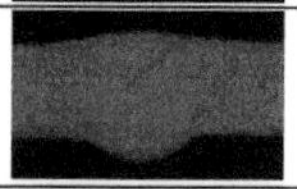
4	800 130 550		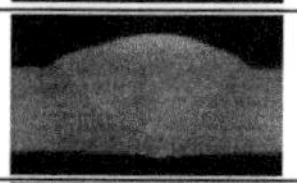
5	800 150 600	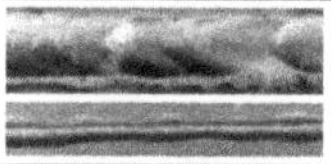	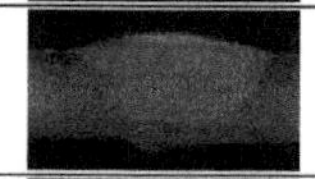
6	800 160 500		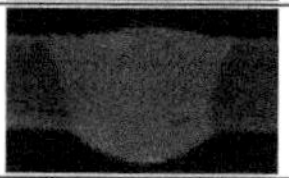
7	900 130 600		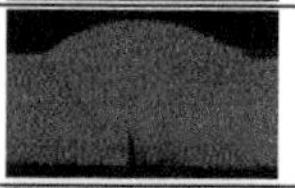
8	900 150 550	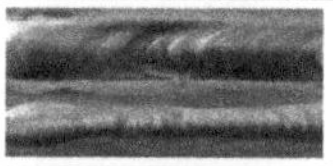	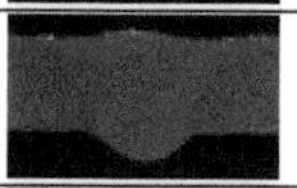
9	900 160 500	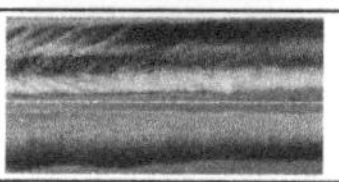	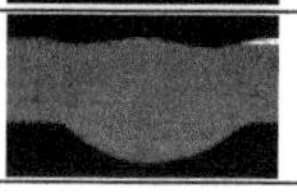

Figure 3. Morphology and cross section of welding joints in different parameters.

It could be found that the arc current has the most significant effect on the weld geometry, compared with the laser beam power and welding speed. Except the arc current, the welding speed was another factor that influenced the reduction of the geometry of the weld seam, which had a significant effect on the front height and back height. Compared with the other two parameters, the laser beam power had a smaller effect on the weld geometry, but it had a greater effect on the weld width, which was higher than the other two factors.

The best parameters of the laser induced arc welding 3 mm thick Al alloys are shown in Table 4. The welding appearance and macrostructure of the joint are shown in Figure 4. The welded joint was obtained in the free formation state. It could be found that the welding appearance was uniform and no hump defects. There was no welding porosity and the cracks in the laser induced arc hybrid welding Al alloys joint. The width of welding beam was about 5 mm on top of the joint and 3 mm at the bottom of the joint. The heat effect zone of the joint was less than 2 mm.

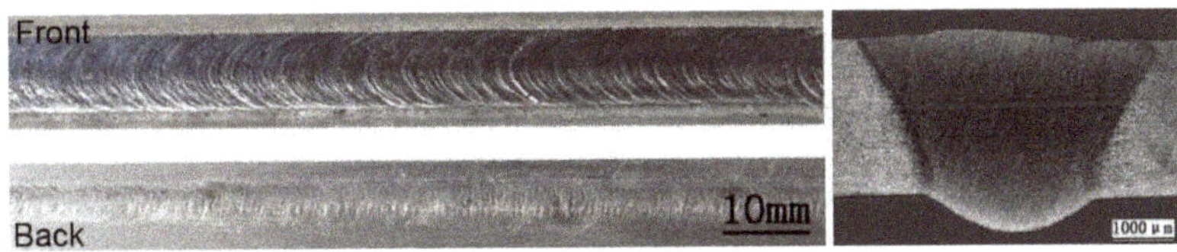

Figure 4. Appearance and the macrostructure of hybrid welding Al alloys joint.

Table 4. Best parameters for 3 mm thick Al alloys hybrid welding process.

Arc Current (A)	Average Laser Power (W)	Welding Speed (mm·Min^{-1})	Pulse Repetition Rate (Hz)	Pulse Peak Power (W)	Filler Speed m/Min
150	900	600	40	8000	2

3.2. Tensile Properties

In order to understand the property of the hybrid welding joint, the tensile tests of the joints were put forward in different conditions. Table 5 shows the results of the transverse tensile test. The base metal (BM) showed a tensile strength and elongation of 340 MPa and 19%, respectively. However, the tensile strength and elongation of the hybrid welding (HW) joint were 234 MPa and 15.2%, respectively.

Table 5. Transverse tensile properties of laser–arc hybrid welded 6061-T6 joints.

Joints	Ultimate Tensile Strength (MPa)	Elongation (%)	Joint Efficiency (%)	Fracture Positions of Tensile Specimens
BM	330~340	18.7~19.4	100	BM
HW	230~240	14.30~15.71	66.8~70.8	HAZ
AW (155 °C, 8 h)	235~245	13.8~16.1	71.2~74.2	HAZ
STA (520 °C)	283~288	19.01~22.02	83.2~87	FZ (Fusion zone)

Firstly, the HW joints were treated with only artificial aging (155 °C, 4/8/12 h), but the only ageing treatment had no obvious improvement in the static tensile properties of the joints, thus it was not discussed.

As it was fractured on the HAZ of the HW joint, the solution treatment and ageing (STA) was done in different conditions. The STA treatment offered the obvious improvement on the mechanical properties of the joints. The values of the tensile strength and elongation of the STA joints were on average 286 MPa and 20.5%, which were 22% and 34% greater than those of HW joints, respectively. It should be pointed out that the fracture locations of HW joints were all in the HAZ, however the solution treated and aged HW joints failed in the fusion zone (FZ). The ratio of the tensile strength of the joints to the tensile strength of base metal was known as the joint efficiency [25–29]. The single

HW joints exhibited joint efficiencies of 68.82%, respectively, while the STA joints showed a high joint efficiency of 83.2%~87%.

3.3. Hardness

The Vickers hardness of the 6061-T6 aluminum alloy was about 110–120 Hv. Due to the heat resource acting on the center of the weld, the heating cycles for both sides of the weld suffered were the same during the welding process, thus the hardness distributions of the two sides should be similar. Figure 5 shows the hardness distribution of the hybrid welded joints under different treatments. The average hardness of the fusion zone was about 60–70 Hv. The lowest the hardness 53 Hv was observed in HAZ. After aging treatment, the hardness of each zone increased a little. However, the location of the lowest hardness was also in the HAZ.

As for the STA treatment joints, the change of the hardness of the fusion zone was similar to that of aged joints and the average hardness was 70 Hv. It was also noteworthy that the hardness of the STA joint had been greatly improved in the heat effect zone. Due to the great change of microstructure, the hardness of the STA treatment BM was lower than that of the original BM.

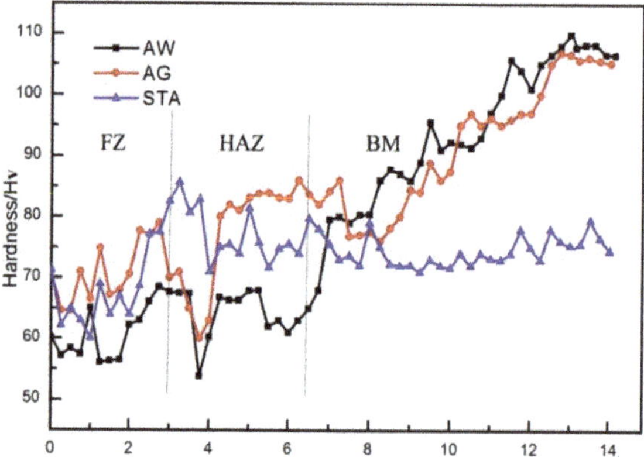

Figure 5. Hardness distributions of different treated joints.

3.4. Microstructure

The hybrid welded joints were initially observed by an optical microscopy and the microstructure of different zones of the HW joint are shown in Figure 6. The joint consists of three zones, namely, a fusion zone (FZ) (Figure 6a), a partial fusion zone (PFZ) (Figure 6b), and a heat effect zone (HAZ) (Figure 6c). The FZ was composed of a large number of equiaxed crystals and a small amount of columnar crystals. In the PFZ, there were plenty of columnar crystals along the direction that was perpendicular to the fusion line. Aluminum alloys have a high thermal conductivity, therefore the welding heat can quickly spread to HAZ, resulting in the microstructure of this zone (Figure 6c) being completely different from that of base metal. The soften area of the joint was formed because of the grain coarsening and the reversion of strengthening β″ (needle-like Mg_2Si) precipitates, thus the tensile tested specimens failed in this region. Plenty of nonhardening β phases (flaky Mg_2Si, black precipitates in Figure 6c) can be found in HAZ, which are formed with the effect of welding thermal cycle.

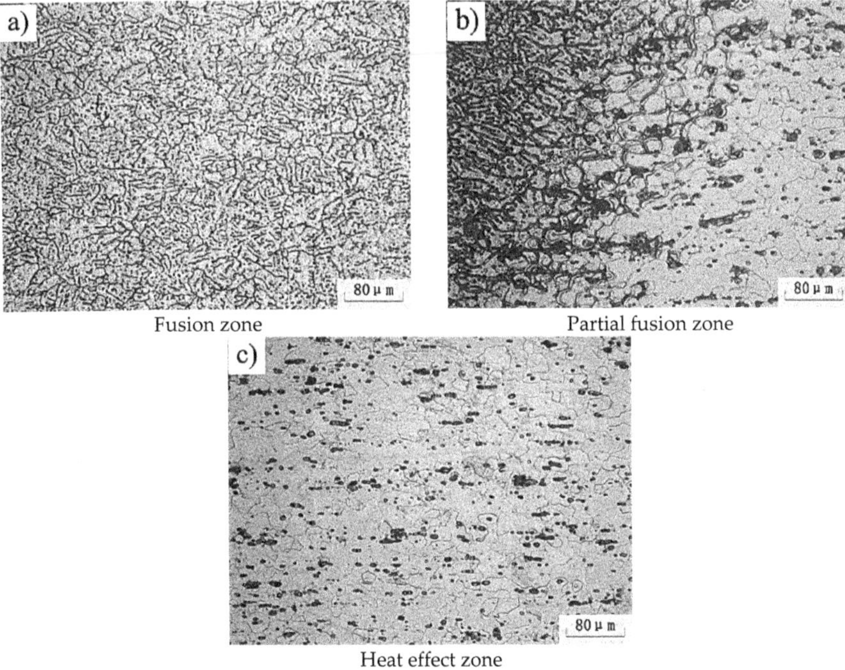

Fusion zone   Partial fusion zone

Heat effect zone

Figure 6. Optical microscopy of hybrid welding 6061-T6 joints.

The different regions of HW joint under STA treatment are presented in Figure 7. Compared with the corresponding area of the as-welded and artificial aged joints, the grain size and morphology had no obvious change, and the number of large size black precipitates was reducing and distributed more uniformly in the α (Al) matrix. However, it did not mean that the microstructure of the HAZ was not changed, as it was not observed simply by the OM test.

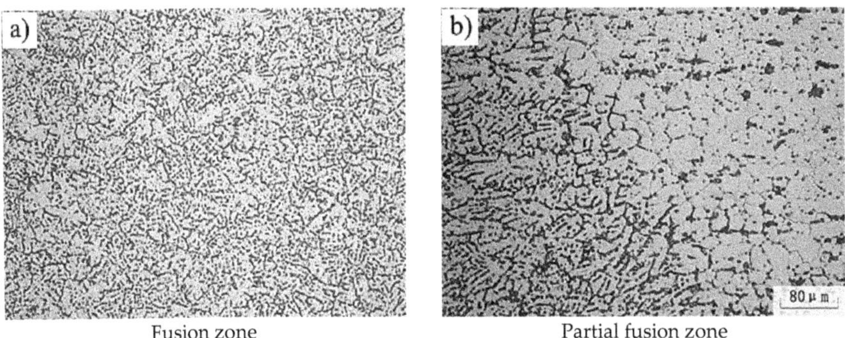

Fusion zone   Partial fusion zone

Figure 7. *Cont.*

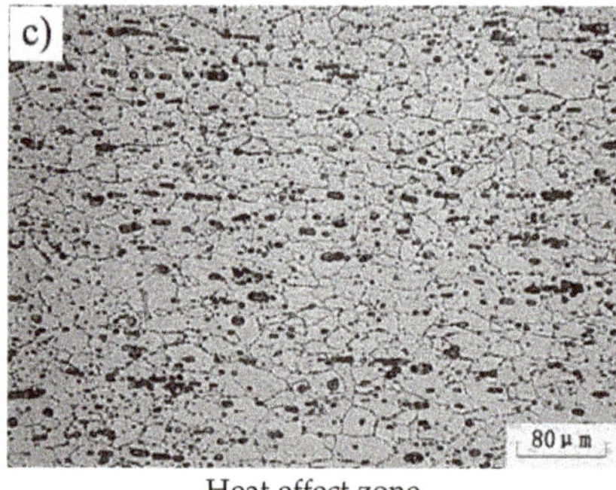

Heat effect zone

Figure 7. Optical microscopy of solution treated and aged 6061-T6 joints.

3.5. Fracture

The failure patterns of the tensile tested specimens are characterized by SEM, as shown in Figure 8. A large number of dimples were found in all the fracture surfaces, which indicated that most of the failure was the result of ductile fracture. It can be observed that the dimples on the fracture surface of the STA joints (Figure 8b) are smaller than those of HW joint (Figure 8a). The experimental results were consistent with the conclusions offered by Hu and Richardson [30], which reported that there were a large number of fine dimples on the surface of the solution heat-treated samples. In the tensile process of ductile material, voids often formed prior to necking. If the necking was generated earlier, the formation of voids would become more dominant. The coarse and elongated dimples could be seen in the welded joints (Figure 8a) [28]. Figure 8c shows the XRD analysis results of the STA joint fracture, and Mg$_2$Si is found on the fracture.

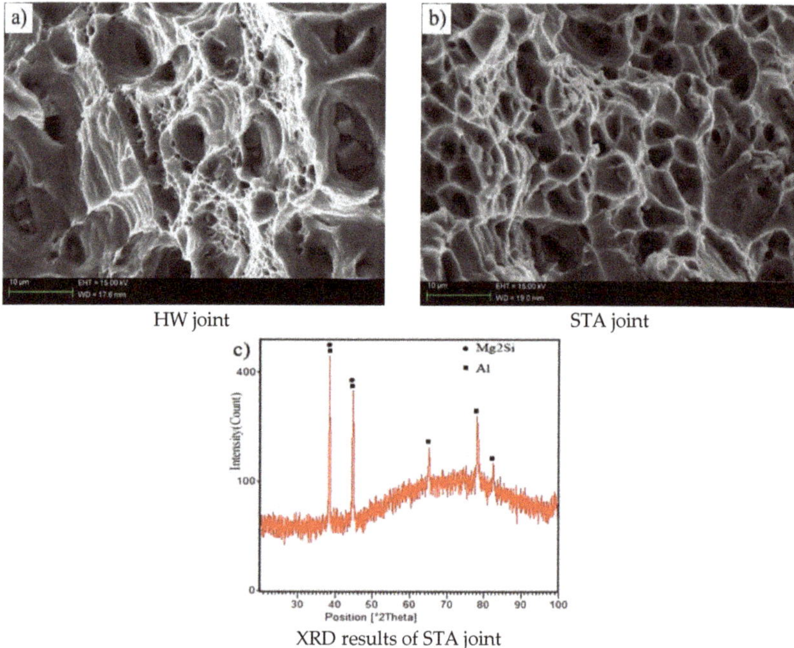

HW joint   STA joint

XRD results of STA joint

Figure 8. SEM and XRD of tensile tested specimen fractures.

3.6. Discussion

Figure 9 shows the distribution and shape of the strengthening precipitates in the FZ and HAZ under different heat treatment conditions. Mg_2Si is the major precipitate in 6000 series aluminum alloys. In the solution and aging (STA) joint, according to the different conditions (temperature), precipitation phases with different morphologies and strengthening effects would be sequentially precipitated. The precipitation process is generally: Supersaturated α solid solution $\rightarrow$ G·PI region (spherical objects without independent lattice structure) $\rightarrow$ G·PII region (needle-like ordered structure β'') $\rightarrow$ β' (rod-like semicoherent structure) $\rightarrow$ β (flaky equilibrium phase Mg_2Si).

The coherent precipitates known as β'' (needle-like Mg_2Si) made the most strength effect on the mechanical property of the BM. A lot of tiny needle-shaped β'' precipitates uniformly distributed through the base metal (Figure 9a,b). In the low-magnification overview, only a small amount of nanometer precipitates could be seen, which illustrated that most of Mg_2Si in the BM exists in the form of β''.

Due to the direct effect of the hybrid welding source and the feeding of ER 5356 filler wire, the component and precipitates of the joints had been modified significantly in the FZ, and the precipitates state had been also changed in the welding process in HAZ. At low magnification, no β'' is found in the FZ of HW joint (Figure 9c). Compared with the base metal, the majority of β'' precipitates are transformed into a weak hardening rod—shaped β' precipitates or the equilibrium flake β phases in the HAZ (Figure 9d), therefore some large particle black β phases can be observed in Figure 6c. As for simple artificial aged (AG) joints, a small quantity of precipitates agglomerated in the FZ (Figure 9e), and the density of the precipitates was higher than that of HW joint HAZ (Figure 9f). In the FZ and HAZ, most of the alloying elements were existing in the equilibrium strengthening phase. Therefore, it was difficult for the precipitation of strengthening β'' phase.

After the HW joint was treated by STA, a lot of nano-sized precipitates can be found in Figure 9e and f. Plenty of β'' were uniformly distributed in the FZ and their average width and length were about 9 and 190 nm, respectively. A great number of precipitates were uniformly dispersed in α (Al) matrix in the HAZ, whose size was varied between 25 and 130 nm. In the hybrid welding process, the Al alloy was fully melted, and the precipitate size was mainly decided by the solidification rate. However, in different parts of the laser induced arc hybrid welding fusion zone, the solidification rate was obviously different, thus the size of precipitate was changed in extent.

In the process of solid solution treatment, the alloy elements were dissolved into α solid solution as much as possible. Therefore, in the subsequent aging process, the precipitates were much more uniformly distributed than that under HW joint or aged state. At the same time, the content of Mg element in feeding of ER5356 filler wire was higher than that of BM, and it would provide a favorable condition for the formation of β'' precipitate, thus the content of precipitates in the FZ was more than that in HAZ.

Figure 10 shows the spectral analysis results of precipitated phases in Figure 9d. Through the analysis of the organization of different regions, it was found that the elemental composition of this region was mainly composed of element Mg and Si. The element Cu was found, because of the influence of copper sample stage. Therefore, it could be deduced that the precipitated phases in this area was mainly the Mg_2Si phase [24].

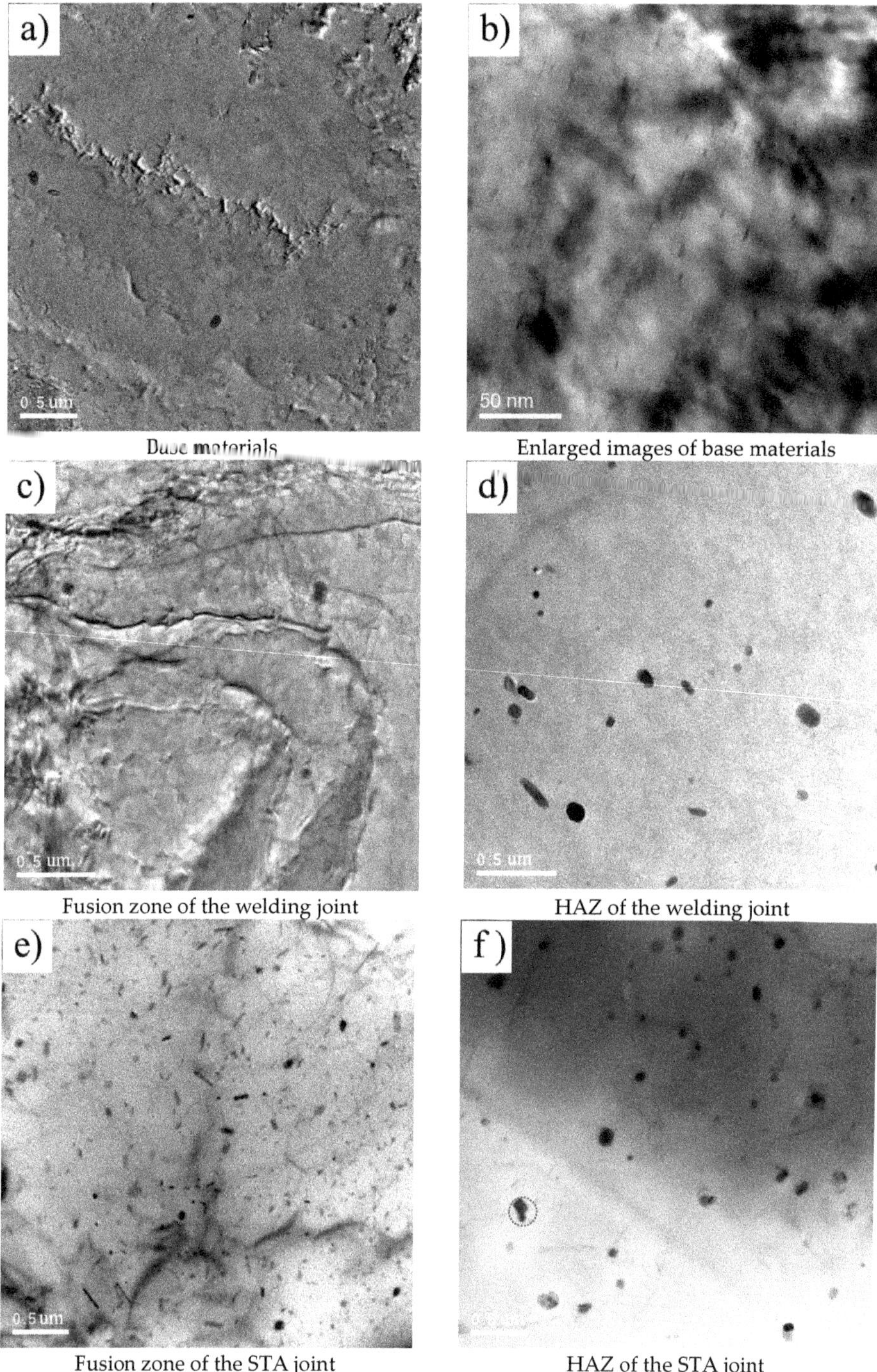

Figure 9. Distribution of β'' and in β in 6061-T6 BM and hybrid welded joints.

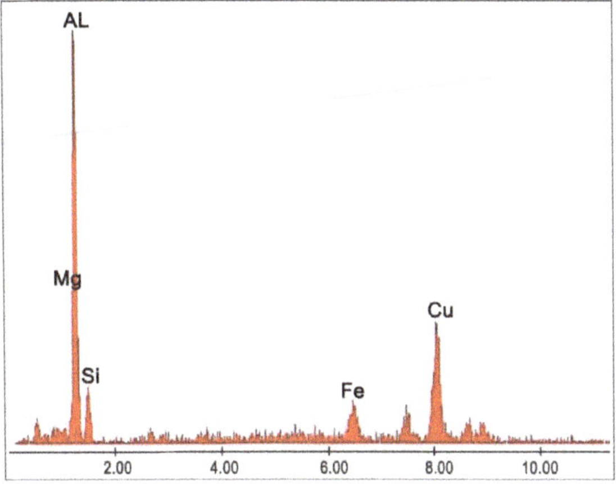

Figure 10. Spectral analysis results of precipitated phases in Figure 9d.

Figure 11 shows the dislocations structures of the BM and HW joints treated by different processes. Since the BM is in the rolling state, the dislocation density inside the BM grains was very high. The density of the dislocation was decrease in the hybrid welding joints (Figure 11a,b), especially in the HAZ of hybrid welding joint, where it did not find the dislocation tangling cells, but instead the dislocation stripes. In STA joints, the dislocation fringes were transformed into a single dislocation fringe with a more uniform distribution (Figure 11d,e). Therefore, the dislocation in STA joint had been rearranged in some extent.

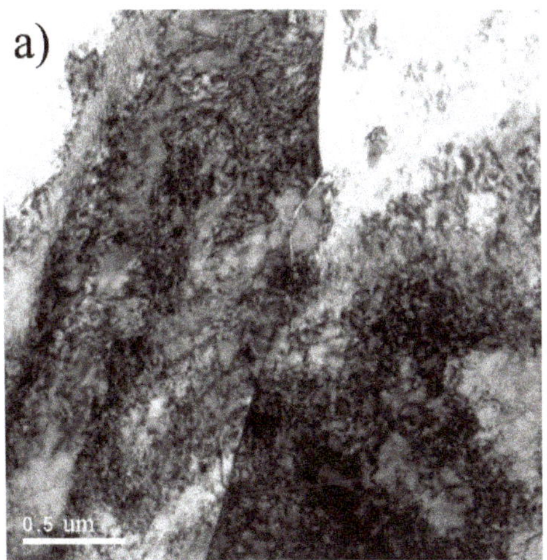

Base metal

Figure 11. *Cont.*

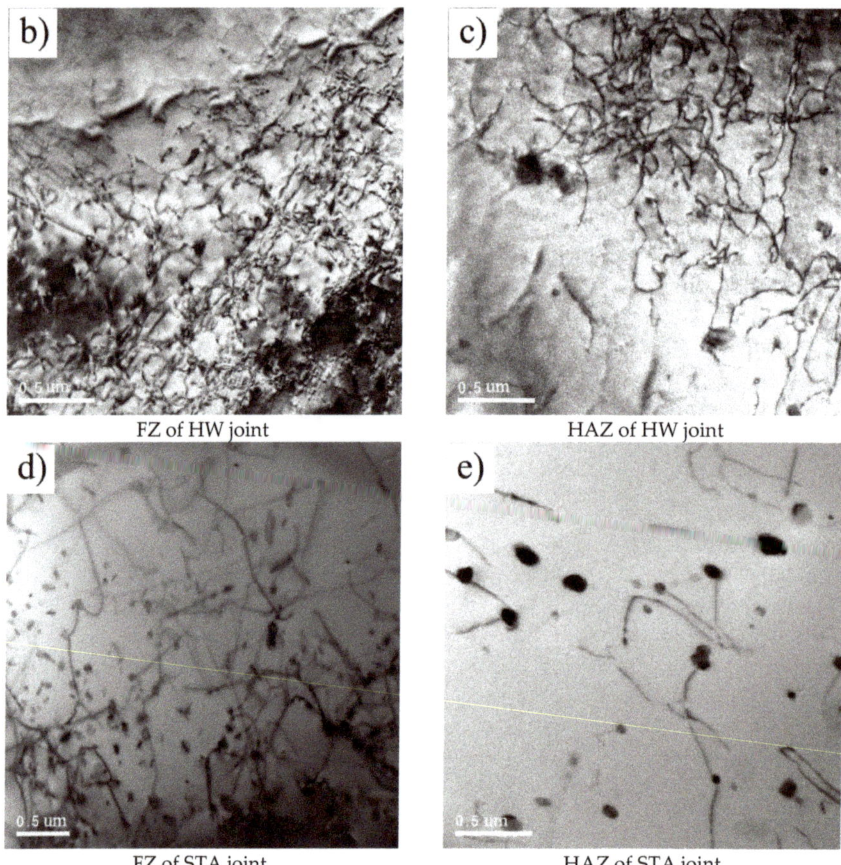

Figure 11. Dislocation structures in 6061-T6 hybrid welded and solution and aging (STA) joints.

From the experimental results, it was found that mechanical properties of HW joints were lower than those of the base metal. The failure location was about 4 mm from the center of the weld, where it corresponded to the position of the lowest hardness. With the effect of welding thermal cycles, the grains were coarse in the HAZ of HW joint (as shown in Figure 6c), and the β'' precipitates transformed into no or weak hardening β' or β phases (Figure 9d). It became the crack sources in the process of tensile tests, therefore, a weak area was formed at the HAZ of HW joint, causing the specimen to failure.

In laser induced TIG hybrid welding Al alloy joint, the welding heat input was obviously lower than that in the TIG welding process. The microstructure in the fusion zone of HW joint was the fine equiaxed grain, whose size was evidently smaller than that in the BM. In the STA process, the grain size of the fusion zone was not changed, which exerted a beneficial effect on the property of the joint. Therefore, it could be deduced that: (1) After STA treatment for HW joint, there is no significant change in grain size in the FZ and it is helpful for the property of the joint; (2) the dislocation configurations in the corresponding regions of the HW joints are different, and the dislocations are changed after STA treatment, which is re-uniformly arranged due to recovery softening; (3) after STA treatment, nano-level strengthening phase β'' with uniform distribution appears in the joint, which plays a dominant role for the property improvement of the STA joint.

In the heat treatment strengthening aluminum alloys, the properties of the joints were determined by the existing forms, quantity, and distribution of the strengthening phases and also related to the density and distribution of the dislocations [31]. The diffuse precipitation precipitates strengthen and the density of dislocations reduction made an obvious influence on the comprehensive properties of materials. In addition, the uneven distribution of dispersed precipitates and dislocations in the HW joint would lead to stress concentration on HAZ in the tensile test process, resulting in uneven deformation of the specimens in each part. Therefore, the properties of the HW joint would be reduced because of the strengthen phase precipitation and stress concentration.

For the STA joints, the dispersion strengthening of β'' phase played a dominant role, thus the tensile strength and ductility of the STA joints were improved. In addition, the dislocations density in the FZ and HAZ of STA joints were changed to a similar level. Still the grain size of the FZ was slightly smaller than that of the base metal, and the grain size of the HAZ was nearly same with the BM. Therefore, the degree of deformation during the tensile test process gradually become uniform, and the elongation of the STA joint has been improved obviously. Due to lower tensile properties of the filler wire, the STA joint all failed approaching the fusion zone.

4. Conclusions

In this research, the influences of welding parameters and STA treatment on the microstructures and mechanical properties of the laser induced TIG hybrid welded 6061-T6 joints have been investigated and the main conclusions can be derived as follows:

The arc current made an obvious effect on the welding joint morphology. The microstructure of the laser induced arc hybrid welding Al alloy joint was mainly composed of the fine equiaxed grain and a small amount of columnar crystals, which made a beneficial effect on the STA treatment process. The aging treatment and solution and aging treatment made little effect on the grain size of the hybrid welded joint.

The tensile property of laser induced arc hybrid welded joint was 69% of the base metal, which was mainly influenced by the reduction of β'' phase. After the solution and aging (STA) treatment, the tensile property of the STA treated joint was 84% of the base metal, and the elongation of the joint was same as the base metal. The tensile strength and the ductility of the laser–arc hybrid welded Al alloy joints were improved by the hybrid effect of grain size refinement in the hybrid welding fusion zone and the STA treatment. After STA treatment, the nano-level strengthening phase β'' with uniform distribution appeared afresh in the joint, which played a dominant role for the property improvement of the joint.

Author Contributions: H.W. Conceptualization, Methodology, Supervision, X.L. Investigation, and L.L. Formal Analysis. All authors contributed to the writing and editing of the manuscript. All authors have read and agreed to the published version of the manuscript.

Funding: The authors gratefully acknowledge the support of the National Natural Science Foundation of China (U1764251 and 51975090).

References

1. Wona, S.; Seoa, B.; Parka, J.M.; Kim, H.K.; Song, K.H.; Min, S.-H.; Ha, T.K.; Park, K. Corrosion behaviors of friction welded dissimilar aluminum alloys. *Mater. Charact.* **2018**, *144*, 652–660. [CrossRef]
2. Monazzah, A.H.; Pouraliakbar, H.; Bagheri, R.; Reihani, S.S. Toughness behavior in roll-bonded laminates based on AA6061/SiCp composites. *Mater. Sci. Eng. A* **2014**, *598*, 162–173. [CrossRef]
3. Monazzah, A.H.; Bagheri, R.; Reihani, S.S.; Pouraliakbar, H. Toughness enhancement in architecturally modified Al6061-5 vol.% SiCp laminated composites. *Int. J. Damage Mech.* **2014**, *24*, 245–262. [CrossRef]
4. Monazzah, A.H.; Pouraliakbar, H.; Bagheri, R.; Reihani, S.M.S. Al-Mg-Si/SiC laminated composites: Fabrication, architectural characteristics, toughness, damage tolerance, fracture mechanisms. *Compos. Part B Eng.* **2017**, *125*, 49–70. [CrossRef]

5.	Pouraliakbar, H.; Nazari, A.; Fataei, P.; Livary, A.K.; Jandaghi, M. Predicting Charpy impact energy of A16061/SiCp laminated nanocomposites in crack divider and crack arrester forms. *Ceram. Int.* **2013**, *39*, 6099–6106. [CrossRef]

6.	Karunakaran, N. Effect of Pulsed Current on Temperature distribution and characteristics of GTA welded magnesium alloy. *IOSR J. Mech. Civ. Eng.* **2013**, *4*, 1–8. [CrossRef]

7.	Najiha, M.S.; Rahman, M.M.; Kamal, M.; Yusoff, A.R.; Kadirgama, K. Minimum quantity lubricant flow analysis in end milling processes: A computational fluid dynamics approach. *J. Mech. Eng. Sci.* **2012**, *3*, 340–345. [CrossRef]

8.	Oliveira, J.P.; Crispim, B.; Zeng, Z.; Omori, T.; Braz Fernandes, F.M.; Miranda, R.M. Microstructure and mechanical properties of gas tungsten arc welded Cu-Al-Mn shape memory alloy rods. *J. Mater. Process. Technol.* **2019**, *271*, 93–100. [CrossRef]

9.	Oliveira, J.P.; Schell, N.; Zhou, N.; Wood, L.; Benafan, O. Laser welding of precipitation strengthened Ni-rich NiTiHf high temperature shape memory alloys: Microstructure and mechanical properties. *Mater. Des.* **2019**, *162*, 229–234. [CrossRef]

10.	Pouraliakbar, H.; Hamedi, M.; Kokabi, A.H.; Nazari, A. Designing of CK45 carbon steel and AISI 304 stainless steel dissimilar welds. *Mater. Res.* **2014**, *17*, 106–114. [CrossRef]

11.	Golezani, A.S.; Barenji, R.V.; Heidarzadeh, A.; Pouraliakbar, H. Elucidating of tool rotational speed in friction stir welding of 7020-T6 aluminum alloy. *Int. J. Adv. Manuf. Technol.* **2015**, *81*, 1155–1164. [CrossRef]

12.	Acherjee, B. Hybrid laser arc welding: State-of-art review. *Opt. Laser Technol.* **2018**, *99*, 60–71. [CrossRef]

13.	Wang, Z.M.; Oliveira, J.P.; Zeng, Z.; Bu, X.Z.; Peng, B.; Shao, X.Y. Laser beam oscillating welding of 5A06 aluminum alloys: Microstructure, porosity and mechanical properties. *Opt. Laser Technol.* **2019**, *111*, 58–65. [CrossRef]

14.	Khorrami, M.S.; Mostafaei, M.A.; Pouraliakbar, H.; Kokabi, A.H. Study on microstructure and mechanical characteristics of low-carbon steel and ferritic stainless steel joints. *Mater. Sci. Eng. A* **2014**, *608*, 35–45. [CrossRef]

15.	David, S.A.; Babu, S.S.; Vitek, J.M. Welding: Solidification and microstructure. *JOM* **2003**, *55*, 14–20. [CrossRef]

16.	Khalaj, G.; Pouraliakbar, H.; Jandaghi, M.R.; Gholami, A. Microalloyed steel welds by HF-ERW technique: Novel PWHT cycles, microstructure evolution and mechanical properties enhancement. *Int. J. Press. Vessel. Pip.* **2017**, *152*, 15–26. [CrossRef]

17.	LH, S.; AR, R. Investigation of aluminum-stainless steel dissimilar weld quality using different filler metals. *Int. J. Automot. Mech. Eng. (IJAME)* **2013**, *8*, 1121–1131.

18.	Ahmad, R.; Bakar, M.A. Effect of a post-weld heat treatment on the mechanical and microstructure properties of AA6061 joints welded by the gas metal arc welding cold metal transfer method. *Mater. Des.* **2011**, *32*, 5120–5126. [CrossRef]

19.	Kulekci, M.K.; Kaluç, E.; Sik, A.; Basturk, O. Experimetal Comparison of MIG and Friction Stir Welding Processes for En AW-6061-T6 (Al Mg-1 Si Cu) Aluminium Alloy. *Arab. J. Sci. Eng.* **2010**, *35*, 322.

20.	Narsimhachary, D.; Bathe, R.N.; Padmanabham, G.; Basu, A. Influence of temperature profile during laser welding of aluminum alloy 6061 T6 on microstructure and mechanical properties. *Mater. Manuf. Process.* **2014**, *29*, 948–953. [CrossRef]

21.	Zhang, C.; Gao, M.; Zeng, X. Effect of microstructural characteristics on high cycle fatigue properties of laser-arc hybrid welded AA6082 aluminum alloy. *J. Mater. Process. Technol.* **2016**, *231*, 479–487. [CrossRef]

22.	Dewan, M.W.; Okeil, A.M. Influence of Weld Defects and Post-Weld Heat Treatment (PWHT) of Gas Tungsten Arc (GTA)-Welded AA-6061-T651 Aluminum Alloy. *J. Manuf. Sci. Eng.* **2015**, *137*. [CrossRef]

23.	Peng, D.; Shen, J.; Tang, Q.; Wu, C.-p.; Zhou, Y.-b. Effects of aging treatment and heat input on the microstructures and mechanical properties of TIG-welded 6061-T6 alloy joints. *Int. J. Miner. Metall. Mater.* **2013**, *20*, 259–265. [CrossRef]

24.	Elangovan, K.; Balasubramanian, V. Influences of post-weld heat treatment on tensile properties of friction stir-welded AA6061 aluminum alloy joints. *Mater. Charact.* **2008**, *59*, 1168–1177. [CrossRef]

25.	Song, G.; Li, T.; Chen, L. Interface bonding mechanism of immiscible Mg/steel by butt fusion welding with filler wire. *Mater. Sci. Eng. A* **2018**, *736*, 306–315. [CrossRef]

26. Jandaghi, M.R.; Pouraliakbar, H. Study on the effect of post-annealing on the microstructural evolutions and mechanical properties of rolled CGPed Aluminum-Manganese-Silicon alloy. *Mater. Sci. Eng. A* **2017**, *679*, 493–503. [CrossRef]

27. Pouraliakbar, H.; Jandaghi, M.R.; Baygi, S.J.M.; Khalaj, G. Microanalysis of crystallographic characteristics and structural transformations in SPDed Al Mn Si alloy by dual-straining. *J. Alloys Compd.* **2017**, *696*, 1189–1198. [CrossRef]

28. Jandaghi, M.R.; Pouraliakbar, H.; Khalaj, G.; Khalaj, M.-J.; Heidarzadeh, A. Study on the post-rolling direction of severely plastic deformed Aluminum-Manganese-Silicon alloy. *Arch. Civ. Mech. Eng.* **2016**, *16*, 876–887. [CrossRef]

29. Myhr, O.R.; Grong, Ø. Process modelling applied to 6082-T6 aluminium weldments—I. Reaction kinetics. *Acta Metall. Mater.* **1991**, *39*, 2693–2702. [CrossRef]

30. Hu, B.; Richardson I, M. Microstructure and mechanical properties of AA7075 (T6) hybrid laser/GMA welds. *Mater. Sci. Eng. A* **2007**, *459*, 94–100. [CrossRef]

31. Moor, P.L.; Howse, D.S.; Wallach, E.R. Development of Nd: YAG laser and laser/MAG hybrid welding for land pipeline application. *Weld. Cut.* **2004**, *3*, 174–180.

Article

Correlations and Scalability of Mechanical Properties on the Micro, Meso and Macro Scale of Precipitation-Hardenable Aluminium Alloy EN AW-6082

Anastasiya Toenjes [1,2,*], **Heike Sonnenberg** [1,2] **and Axel von Hehl** [1,2,3]

[1] Leibniz Institute for Materials Engineering—IWT, Badgasteiner Str. 3, 28359 Bremen, Germany; sonnenberg@iwt-bremen.de (H.S.); vonhehl@iwt-bremen.de (A.v.H.)
[2] Faculty of Production Engineering, University of Bremen, Badgasteiner Straße 1, 28359 Bremen, Germany
[3] MAPEX Center for Materials and Processes, University of Bremen, Bibliothekstr. 1, 28359 Bremen, Germany
* Correspondence: toenjes@iwt-bremen.de; Tel.: +49-421-218-51491

Received: 7 April 2020; Accepted: 6 May 2020; Published: 8 May 2020

Abstract: The mechanical properties of heat-treatable aluminium alloys are improved and adjusted by three different heat treatment steps, which include solution annealing, quenching, and aging. Due to metal-physical correlations, variations in heat treatment temperatures and times lead to different microstructural conditions with differences in the size and number of phases and their volume fraction in the microstructure. In this work, the investigations of the correlation between microhardness measurements on micro samples and the conventional mechanical properties (hardness, yield strength and tensile strength) of macro samples and the comparability of the different heat treatment states of micro and macro samples made of a hardenable aluminium alloy EN AW-6082 will be discussed. Using the correlations between the mechanical properties of micro samples and macro samples, the size of the samples and, thus, the testing cost and effort can be reduced.

Keywords: characterization methods; scalability; mechanical properties; aluminium alloys

---

1. Introduction

The mechanical properties of heat-treatable aluminium alloys are improved and adjusted by three different heat treatment steps, which include solution annealing, quenching and aging. Due to metal-physical correlations, variations in heat treatment temperatures and times lead to different microstructural conditions with differences in the size, the number of phases and the volume fractions in the microstructure. The resulting mechanical performance depends on the heat treatment parameters, and the material condition can be described in relation to the most often-preferred T6-condition, which reaches maximum hardness and strength values.

In the case that the mechanical properties of a new alloy need to be estimated fast, either modeling or experimental high-throughput characterization methods can be used. Most models for calculating the mechanical properties of precipitation-hardenable aluminium alloys after the heat treatment are based on metal-physical interrelations and require exact knowledge of the chemical composition of the alloy and the material conditions. The basis for calculating the changes in mechanical properties is information about the microstructure development of the heat-treated parts during the precipitation hardening. Therefore, many models require the parameters of the strength-enhancing phases such as size, number and/or distribution as input variables for the prediction of mechanical properties. The determination of the input variables requires experimental methods with high technical effort and specific sample preparation, such as small-angle X-ray scattering [1,2], small-angle neutron scattering [3], transmission electron microscopy [2–6] or differential scanning calorimetry [5,7]. In many cases, the basic equations of solid solution strengthening and precipitation hardening are used to calculate such mechanical properties as the yield, tensile strength and hardness [4,5,7–9]. As shown in Reference [10], the correlation between the Vickers hardness and tensile strength for the same material can also be determined accurately using a relatively simple relationship that depends on the quenching rate, aging temperature and aging time.

If the properties need to be characterized in a high-throughput method, the speed of both sample production and characterization must be increased. One way to reduce the processing time of different melts is to reduce the sample size. However, rapid material development means not only the rapid production of new alloys, such as rapid alloy prototyping [11], but also rapid material characterization [12]. A disadvantage of this is that many test methods such as tensile tests require a specific specimen geometry and size.

Within the Collaborative Research Center "Farbige Zustände" (CRC 1232) of the German Research Foundation (DFG), a novel high-throughput method for evolutionary structural material development is approached using micro samples (<1 mm) and adapted mechanical testing methods, which, therefore, reduces the testing cost and effort [13]. Microstructural changes as well as manufacturing influences on the material state of steel and aluminium materials are investigated. Spherical micro samples of different alloy compositions can be provided very quickly with droplet generators, as described in Reference [14], which enables both time-reduced and cost-reduced manufacturing of new types of alloys by using small batches. Many characterization methods of these samples have already been presented in Reference [15]. By combining results of different short-term characterization methods on micro samples such as the micro compression test [16], the potential of transferability from the micro to the macro scale regarding diverse material properties as strength, ductility and even machinability is increased.

The aim of this work is to investigate the correlation between microhardness measurements on micro samples and conventional mechanical properties, e.g., the tensile strength of macro samples, to reduce the testing cost and effort on the example of the conventionally used and well-known precipitation-hardenable aluminium alloy EN AW-6082. The scalability is investigated with the help of mid-range sized meso samples on which conventional testing (HV5 and tensile tests) can take place. Therefore, a direct connection can be drawn from the meso sample to the macro sample as well as from the meso sample to the micro sample since the universal microhardness (UMH) measurements can be performed on the micro level.

2. Materials and Methods

2.1. Material and Sample Geometry

All experimental investigations were performed on the aluminium alloy EN AW-6082. The chemical composition of the alloy was measured by glow discharge optical emission spectroscopy (GDOES) (MICA Analysen GmbH, Düsseldorf, Germany) and is presented in Table 1.

Table 1. Chemical analysis of the samples in wt.% in comparison with the limitations of the DIN EN 573-3:2013-12.

Material		Chemical Composition in wt.%						
		Al	Si	Mg	Mn	Fe	Cu	Others
Samples alloy		bal.	0.916	0.727	0.437	0.381	0.086	<0.05
DIN EN 573-3:2013-12	min.	bal.	0.7	0.6	0.40	-	-	-
	max.		1.3	1.2	1.0	0.50	0.10	0.15

In this study, three different sample geometries are used (Figure 1): conventional tensile specimens, referred to as macro samples (a), flat specimens with a thickness of 2.5 mm, referred to as meso samples (b) and spherical specimens with a diameter of 1 mm, referred to as micro samples (c).

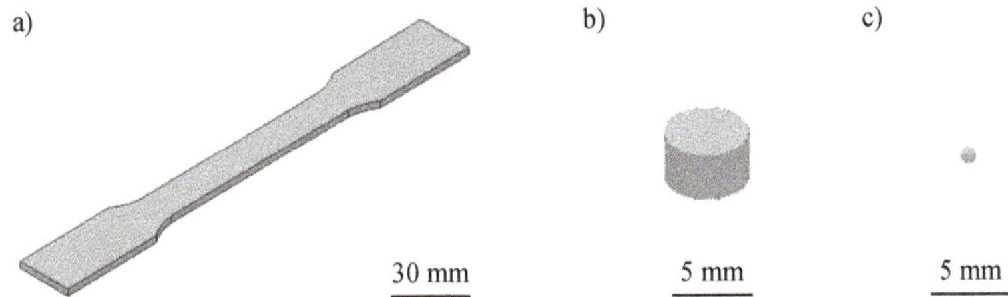

Figure 1. (**a**) Macro, (**b**) meso and (**c**) micro samples.

The flat tensile specimens with a length of 170 mm and small round samples with a diameter of 4 mm were both manufactured from rolled sheets in a T6 condition with a thickness of 2.5 mm. The spherical micro samples were produced in a droplet generator. The micro cross-section of the samples before heat treatment can be seen in Figure 2. Figure 2a,b show a stretched microstructure typical for rolled sheets with more stretched grains in rolling direction. The micro samples show both a cellular and dendritic microstructure (Figure 2c).

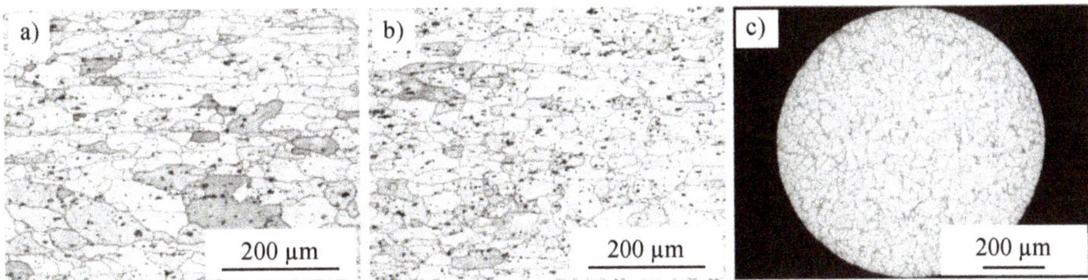

Figure 2. Micro cross-section of the macro and meso samples before the heat treatment in (**a**) a longitudinal to rolling direction, (**b**) transverse to rolling direction and (**c**) of the micro samples.

Due to the small sample size, micro and meso samples were prepared by embedding in a cold curing epoxy resin and grinding to an equatorial plane with water-cooled SiC abrasive paper (grit size of 320 to 1200). Before carrying out the tests, the surface of the samples was polished with a 1 µm diamond suspension (roughness S_a of 4–8 nm).

2.2. Heat Treatment

In order to adjust the mechanical properties, precipitation hardening was applied. The design of the experiments was conducted using Cornerstone® software (version 7.1.0.2, camLine GmbH, Petershausen, Germany). The variation of heat treatment parameters was carried out according to the D-Optimal experimental design [17]. The advantages of a D-Optimal experimental design included significantly reduced experimental effort and maximum prediction power in the selected range of the parameters. The parameters of solution annealing temperature (T_s) and times (t_s) as well as aging temperature (T_a) and duration (t_a) were varied. Air, water and a polymer solution were used as quenching media (Table 2). With these three quenching media, slow cooling, fast quenching and a fast quenching with a reduction of the Leidenfrost [18] phenomenon could be achieved. During quenching in water, the samples were immersed in a circulated water bath at a temperature of approximately 20 °C. Polymer quenching took place in a similar way to water quenching. During air cooling, the samples were slowly cooled with an air blower.

Table 2. Heat treatment parameters.

No.	T_s (°C)	t_s (h)	QM	T_a (°C)	t_a (h)
1	500	0.25	Water	100	20
2	500	0.25	Water	240	2
3	500	0.25	Gas	100	2
4	500	0.25	Gas	170	11
5	500	0.25	Gas	240	20
6	500	0.25	Polymer	240	20
7	500	2.13	Polymer	100	2
8	500	4	Water	100	2
9	500	4	Water	240	20
10	500	4	Gas	100	20
11	500	4	Gas	240	2
12	500	4	Polymer	100	20
13	500	4	Polymer	170	2
14	530	0.25	Polymer	100	11
15	530	2.13	Gas	170	20
16	530	4	Gas	100	2
17	530	4	Polymer	240	2
18	560	0.25	Water	100	2
19	560	0.25	Water	240	20
20	560	0.25	Gas	100	20
21	560	0.25	Gas	240	2
22	560	0.25	Polymer	170	20
23	560	0.25	Polymer	240	11
24	560	2.13	Polymer	240	11
25	560	4	Water	100	20
26	560	4	Water	240	2
27	560	4	Gas	100	11
28	560	4	Gas	240	20
29	560	4	Polymer	100	2
30	560	4	Polymer	240	20

With the selected parameters not only T6, but also overaged and underaged heat treatment conditions were reached. All these heat treatment states were characterized in Reference [10] with differential scanning calorimetry.

2.3. Experimental Testing

After the heat treatment of the samples, different experimental tests were carried out. In quasi-static tensile tests, the tensile strength (R_m) and yield strength ($R_{p0.2}$) were measured on the macro samples. On both macro and meso samples, Vickers hardness measurements (HV5) were carried out. Universal micro hardness (UMH) measurements were conducted on meso and micro samples.

For the quasi-static tensile tests of the macro samples, a servo-hydraulic testing machine of the type PC 160M (Schenck, Darmstadt, Germany) was used. The tests were carried out according to DIN EN 6892-1 at a displacement rate of 1 mm/min using strain gauges of the type FCA-2-23 (Preusser Messtechnik, Bergisch Gladbach, Germany) to determine the yield strength $R_{p0.2}$ at a 0.2% strain. Three macro samples of the same heat treatment condition were examined.

The Vickers hardness measurements were conducted according to DIN EN ISO 6507-1 on a hardness testing device of the type LV-700AT (LECO Corporation, St. Joseph, MI, USA) with a load of 5 kp (HV5). For the Vickers hardness testing, the repetition of the experiment varied due to the sample size and required spacings of indentations. There were five hardness measurements done on each macro sample (N = 5) and only one measurement on each meso sample (N = 1) since these meso samples were too small for further indentations. However, three macro and meso samples of the same heat treatment condition were used to calculate the mean values.

The universal hardness measurements were performed on a Fischerscope H100C testing device (Helmut Fischer GmbH, Sindelfingen, Germany) using a Vickers indenter and a maximum indentation force of 200 mN. The loading time, holding time and relief time were set to 10 s each. There were 25 individual measurements carried out on each sample while investigating three meso and micro samples in each heat treatment condition. Therefore, there were 75 measurements in total for each condition. In the following, the indentation hardness (HIT 0.2/10/10/10) is used as a characteristic value of UMH measurements for further comparison.

3. Results and Discussion

The aim of this work is to find correlations between mechanical properties of micro samples and macro samples. The scalability is investigated with the help of mid-range sized meso samples on which conventional hardness testing (HV5) can take place. Therefore, a direct connection can be drawn from the meso sample to macro sample as well as from the meso sample to micro sample since the UMH measurements can be performed on a micro sample. Figure 3 shows the measured hardness of macro samples when compared to the hardness of meso samples. Only those heat treatment states were further investigated, which exhibited a hardness deviation between the macro samples and meso samples of less than ±10%.

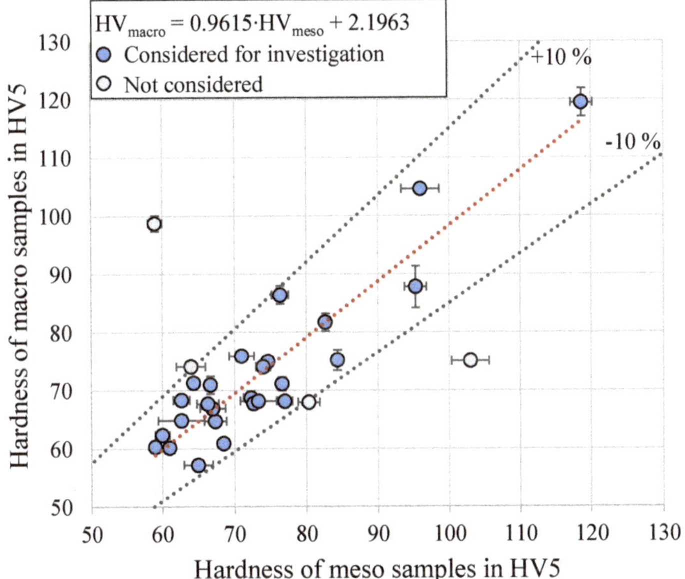

Figure 3. Correlation between hardness of macro and meso samples.

In order to determine the correlation between macro and micro samples, the correlation between the hardness and indentation hardness of meso samples is first investigated (Figure 4).

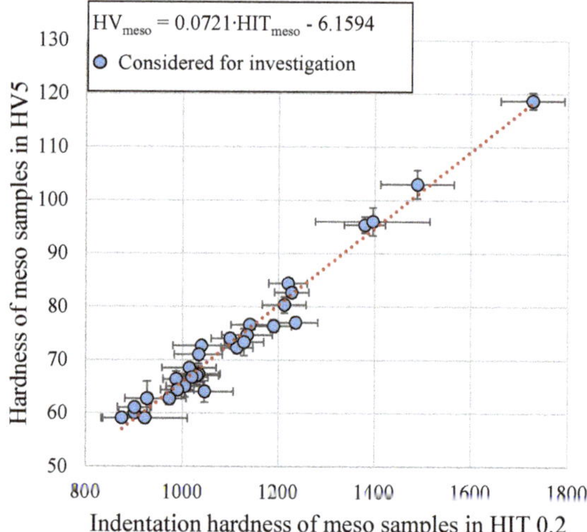

Figure 4. Correlation between the hardness and indentation hardness of meso samples.

The correlation between the indentation hardness of meso and micro samples is shown in Figure 5. In the same way as for the hardness of macro and meso samples, only those heat treatment states were further investigated, which had a hardness deviation between the meso and micro samples of less than ±10%.

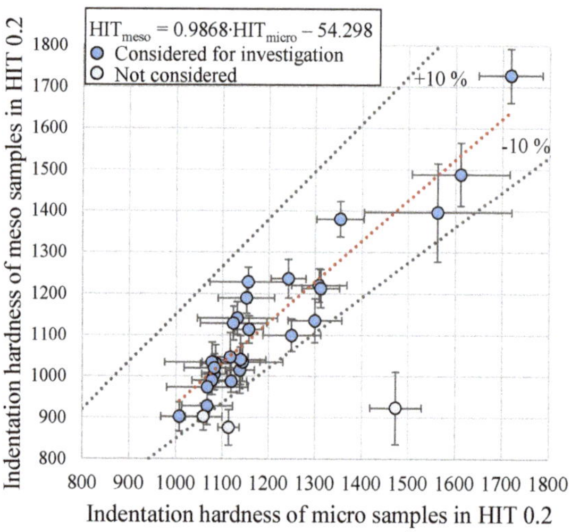

Figure 5. Correlation between indentation hardness of meso and micro samples.

Thus, the correlations between the indentation hardness of the micro samples and the indentation hardness of the meso samples, correlation between the indentation hardness of the meso samples and the hardness of the meso samples and correlation between the hardness of the meso samples and the hardness of the macro samples were established (Figure 6).

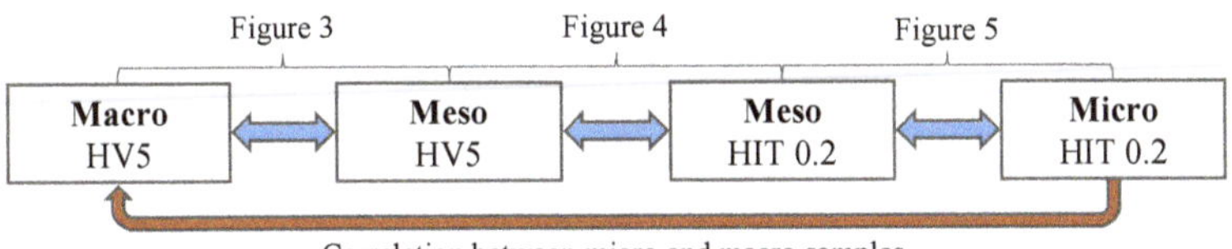

Figure 6. Schematic overview on investigated correlations between micro, meso and macro samples based on conventional hardness (HV5) and indentation hardness (HIT 0.2).

The equation for calculating the hardness of macro samples using the measurements of the indentation hardness on the micro samples is, therefore, as follows.

$$HV_{macro} = 0.0683 \cdot HIT_{micro} - 7.4901 \tag{1}$$

The hardness calculated with Equation (1) as well as the measured hardness are summarized in Figure 7. Although the sample size and microstructure were different, a good accordance between the calculated and measured hardness was achieved. The deviations were due to the individual measuring methods on the one hand and to the differences in the heat treatment conditions on the other hand. The macro, meso and micro samples were heat-treated with the same parameters, but the small samples cooled down faster when compared to the large samples, which may have affected the final properties. The data points excluded in Figures 3 and 5 were not considered in the calculation of the coefficient of determination.

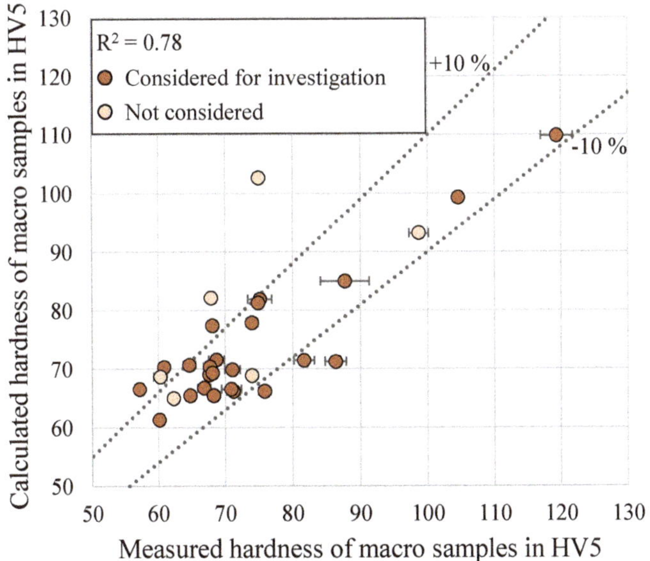

Figure 7. Calculated and experimental data for the hardness of macro samples.

The correlation between hardness and tensile strength as well as the yield strength can be seen in Figures 8 and 9.

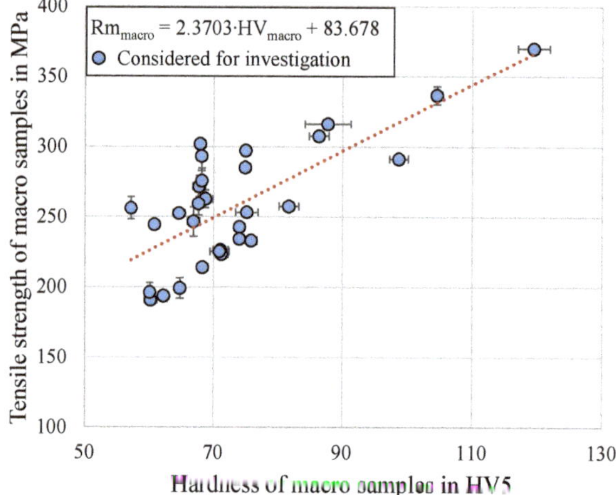

Figure 8. Correlation between tensile strength and hardness of macro samples.

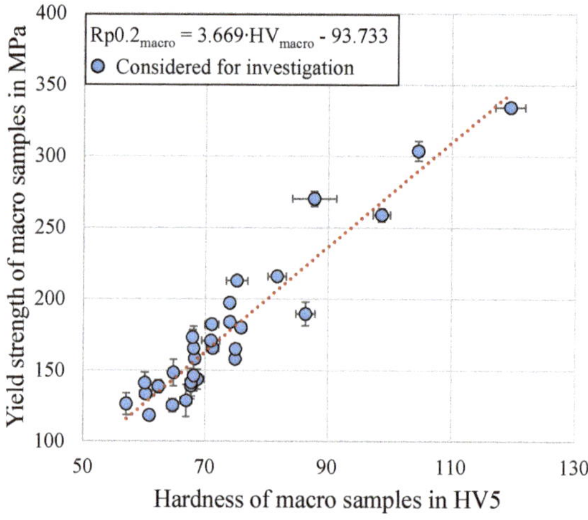

Figure 9. Correlation between yield strength and hardness of macro samples.

The tensile strength and yield strength were calculated with the correlations determined in Figures 8 and 9. The hardness was calculated using Equation (1). The results were summarized in Figures 10 and 11. It was, thus, possible to demonstrate that the tensile strength and the yield strength can be predicted based on the indentation hardness, measured on micro samples, even though the macro and micro samples were manufactured in a different way with resulting differences in the microstructure. In the same way as for the calculation of hardness, the data points excluded in Figures 3 and 5 were not considered in the calculation of the coefficient of determination.

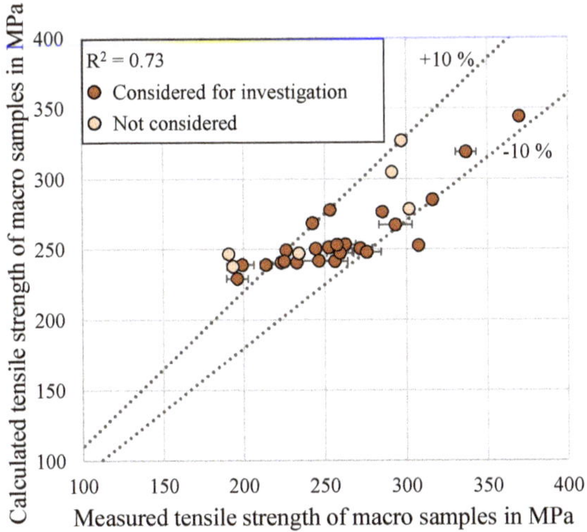

Figure 10. Calculated and experimental data for the tensile strength of macro samples.

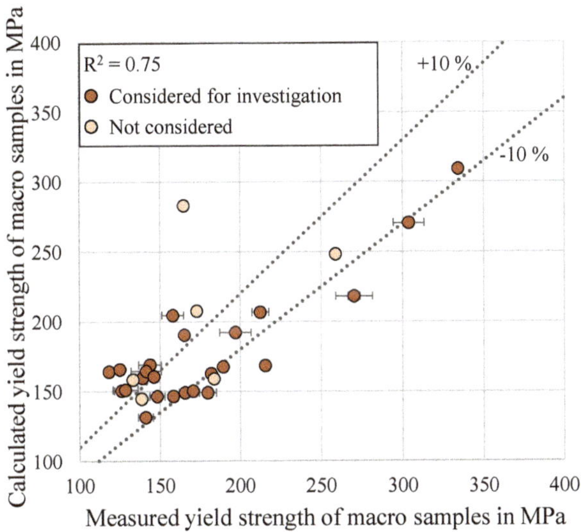

Figure 11. Calculated and experimental data for the yield strength of macro samples.

4. Conclusions

In this paper, a method for the fast and relatively accurate calculation of mechanical properties after heat treatment (i.e., solution annealing, quenching and ageing) was presented using the example of an age-hardenable aluminium alloy based on indentation hardness measurements on micro samples. Using mid-range sized meso samples on which conventional hardness (HV5) and indentation hardness measurements could be performed, a connection between the results of micro samples and macro samples could be established, which allows a direct investigation of the scalability. It was demonstrated that, despite different production routes and microstructures, the micro samples have a similar hardness to the meso and macro samples.

To evaluate this approach, the coefficient of determination was calculated for the correlation between hardness (HV5) and indentation hardness (HIT 0.2/10/10/10) on a micro scale, which results in $R^2 = 0.78$. Furthermore, a coefficient of determination of 0.73 for the tensile strength and 0.75 for the yield strength was calculated. Therefore, the results of this calculation method on the example of the aluminium alloy EN AW-6082 show the potential of predicting mechanical properties by using fast and resource-efficient producible micro samples. The transfer of this method to other alloys remains to be evaluated.

Author Contributions: A.T. performed all experiments, analyzed the obtained data, and wrote the paper with the aid of H.S. in questions of UMH. A.v.H. supervised all investigations, and supported the work with his scientific knowledge. All authors have read and agreed to the published version of the manuscript.

Funding: This research was funded by The German Research Foundation (DFG), Deutsche Forschungsgemeinschaft, grant number: 276397488.

Acknowledgments: Financial support of subprojects U03 "Thermal and thermomechanical heat treatment" and D01 "Qualification of material conditions with mechanical and physical measuring methods" of the Collaborative Research Center CRC 1232 by the Deutsche Forschungsgemeinschaft (DFG, German Research Foundation)-project number 276397488-SFB 1232 is gratefully acknowledged.

Conflicts of Interest: The authors declare no conflict of interest.

References

1. Nicolas, M.; Deschamps, A. Characterisation and modelling of precipitate evolution in an Al-Zn-Mg alloy during non-isothermal heat treatments. *Acta Mater.* **2003**, *51*, 6077–6094. [CrossRef]
2. Werenskiold, J.C.; Deschamps, A.; Brechet, Y. Characterization and modeling of precipitation kinetics in an Al-Zn-Mg alloy. *Mater. Sci. Eng.* **2000**, *A293*, 267–274. [CrossRef]
3. Bardel, D.; Perez, M.; Nelias, D.; Deschamps, A.; Hutchinson, C.R.; Maisonnette, D.; Chaise, T.; Garnier, J.; Bourlier, F. Coupled precipitation and yield strength modelling for non-isothermal treatments of a 6061 aluminium alloy. *Acta Mater.* **2014**, *62*, 129–140. [CrossRef]
4. Esmaeili, S.; Lloyd, D.J. Modeling of precipitation hardening in pre-aged AlMgSi(Cu) alloys. *Acta Mater.* **2005**, *53*, 5257–5271. [CrossRef]
5. Khan, I.N.; Starink, M.J.; Yan, J.L. A model for precipitation kinetics and strengthening in Al-Cu-Mg alloys. *Mater. Sci. Eng.* **2008**, *A472*, 66–74. [CrossRef]
6. Österreicher, J.A.; Papenberg, N.P.; Kumar, M.; Ma, D.; Schwarz, S.; Schlögl, C.M. Quantitative prediction of the mechanical properties of precipitation-hardened alloys with special application to Al-Mg-Si. *Mater. Sci. Eng. A* **2017**, *703*, 380–385. [CrossRef]
7. Sepehrband, P.; Esmaeili, S. Application of recently developed approaches to microstructural characterization and yield strength modeling of aluminum alloy AA7030. *Mater. Sci. Eng. A* **2008**, *487*, 309–315. [CrossRef]
8. Esmaeili, S.; Lloyd, D.J.; Poole, W.J. A yield strength model for the Al-Mg-Si-Cu alloy AA6111. *Acta Mater.* **2003**, *51*, 2243–2257. [CrossRef]
9. Zander, J.; Sandström, R. One parameter model for strength properties of hardenable aluminium alloys. *Mater. Des.* **2008**, *29*, 1540–1548. [CrossRef]
10. Toenjes, A. Empirische Methode zur schnellen Charakterisierung von Wärmebehandlungszuständen hochfester Aluminiumlegierungen. Ph.D. Thesis, Universität Bremen, Bremen, Germany, 2019.
11. Springer, H.; Raabe, D. Rapid alloy prototyping: Compositional and thermo-mechanical high throughput bulk combinatorial design of structural materials based on the example of 30Mn-1.2C-xAl triplex steels. *Acta Mater.* **2012**, *60*, 4950–4959. [CrossRef]
12. Potyrailo, R.; Rajan, K.; Stoewe, K.; Takeuchi, I.; Chisholm, B.; Lam, H. Combinatorial and high-throughput screening of materials libraries: Review of state of the art. *ACS Combinatorial Sci.* **2011**, *13*, 579–633. [CrossRef] [PubMed]
13. Ellendt, N.; Mädler, L. High-Throughput Exploration of Evolutionary Structural Materials. *HTM J. Heat Treat. Mater.* **2018**, *73*, 3–12. [CrossRef]
14. Moqadam, S.I.; Madler, L.; Ellendt, N. Microstructure Adjustment of Spherical Micro-samples for High-Throughput Analysis Using a Drop-on-Demand Droplet Generator. *Materials* **2019**, *12*, 3769. [CrossRef] [PubMed]

15. Steinbacher, M.; Alexe, G.; Baune, M.; Bobrov, I.; Bösing, I.; Clausen, B.; Czotscher, T.; Riemer, O.; Sonnenberg, H.; Thomann, A.; et al. Descriptors for High Throughput in Structural Materials Development. *High-throughput* **2019**, *8*, 22. [CrossRef] [PubMed]
16. Sonnenberg, H.; Clausen, B. Short-Term Characterization of Spherical 100Cr6 Steel Samples Using Micro Compression Test. *Materials* **2020**, *13*, 733. [CrossRef] [PubMed]
17. Box, G.E.P.; Lucas, H.L. Design of experiments in nonlinear situations. *Biometrika* **1959**, *46*, 77–90. [CrossRef]
18. Sarmiento, G.; Bronzini, C.; Canale, A.; Canale, L.; Totten, G. Water and Polymer Quenching of Aluminum Alloys: A Review of the Effect of Surface Condition, Water Temperature, and Polymer Quenchant Concentration on the Yield Strength of 7075-T6 Aluminum Plate. *J. ASTM Int.* **2008**, *6*, 1–18.

MDPI
St. Alban-Anlage 66
4052 Basel
Switzerland
Tel. +41 61 683 77 34
Fax +41 61 302 89 18
www.mdpi.com

Metals Editorial Office
E-mail: metals@mdpi.com
www.mdpi.com/journal/metals